ALGEBRA
ESSENTIALS

ALGEBRA ESSENTIALS

David A. Santos

Olgha B. Davis

MERCURY LEARNING AND INFORMATION

Dulles, Virginia
Boston, Massachusetts
New Delhi

Publisher: David Pallai
MERCURY LEARNING AND INFORMATION
22841 QUICKSILVER DRIVE
Dulles, VA 20166
info@merclearning.com
www.merclearning.com
(800) 232-0223

David A. Santos and Olgha B. Davis. *Algebra Essentials.*
ISBN: 978-1-937585-22-8

The publisher recognizes and respects all marks used by companies, manufacturers, and developers as a means to distinguish their products. All brand names and product names mentioned in this book are trademarks or service marks of their respective companies. Any omission or misuse (of any kind) of service marks or trademarks, etc. is not an attempt to infringe on the property of others.

Library of Congress Control Number: 2015934531
161718321 This book is printed on acid-free paper.

Our titles are available for adoption, license, or bulk purchase by institutions, corporations, etc.
For additional information, please contact the Customer Service Dept. at 800-232-0223(toll free).

All of our titles are available in digital format at authorcloudware.com and other digital vendors. All issues regarding this title can be addressed by contacting info@merclearning.com. The sole obligation of MERCURY LEARNING AND INFORMATION to the purchaser is to replace the book, based on defective materials or faulty workmanship, but not based on the operation or functionality of the product. *Files on the companion disc are also available by writing to the publisher at info@merclearning.com.*

CONTENTS

PREFACE

Welcome to *Algebra Essentials*! This book is designed for instructors to introduce algebra to students in approachable and meaningful ways. *Algebra Essentials* is intended to increase your knowledge and skills needed to succeed in collegiate mathematics courses. The eight chapters in this book present many examples and exercises—with complete solutions—that can be used to practice algebraic concepts. You will receive the most benefit from this book if you use all of the components included. For self-correcting exercises, figures from the text (including 4-color), theorem proofs, and additional information, check out the companion disc. Additional instructor supplements for use as a textbook are available from the publisher.

Exercises

This self-study text contains hundreds of exercises to supplement the numerous examples and aid in the drill and mastery of the concepts:

1) *Supplementary Exercises (End of each section)*

These are found at the end of *each section* of every chapter (e.g., after Sections 2.1, 2.2, 2.3, etc.)

Answers to *ALL* of these exercises may be found in *Appendix B* at the end of the book.

2) *Review Exercises (End of the book)*

Over 100 of these exercises (included to reinforce or review concepts) may be found in

Appendix A. Answers to the *odd-numbered* Review Exercises may be found in *Appendix C*.

3) *Self-correcting Mini-Quizzes*

The companion disc in the back of the book includes seven, multiple-choice, self-correcting mini-quizzes on arithmetic, equations, factoring, inequalities, like-terms, exponents, and word problems.

Companion files on the disc are also available by writing to the publisher at info@merclearning.com.

ACKNOWLEDGMENTS

Many people have contributed to this work in very significant ways. First, we would like to thank Margaret Hitczenko, John Majewicz, and José Mason for their numerous suggestions and ideas. We also would like to thank Iraj Kalantari, Lasse Skov, and Don Stalk for their editorial suggestions.

It is with great humility and gratitude that Olgha Davis thanks Tuere Bowles, Christine Grants, David Keim, and Roberta Mahatha, who were instrumental in providing support and insightful guidance throughout this project. And last, but far from least, Olgha Davis could not have completed this book without her two beautiful daughters, Dalya and Andreya Davis.

David A. Santos
Olgha B. Davis

FIRST IMPRESSIONS

CHAPTER 1

WHY STUDY ALGEBRA?

This chapter is optional. However, it is necessary to answer two commonly asked questions about learning algebra: Why do adult learners need to learn algebra? And when will they use it in the real world? We attempt to answer these questions in the *Value in Studying Algebra*, *What Is Elementary Algebra All About?* and the *Puzzles* sections.

1.1 THE VALUE IN STUDYING ALGEBRA

Mathematicians like to boast about the multiple applications of their discipline to real-world problems, as commonly found in science, business, engineering, etc. For example, chemical compounds must obey certain geometric arrangements, which in turn specify how they behave. Financial firms use mathematical theory to explain long-term markets behavior. Biomedical engineering researchers use mathematics to model the way virus strands twist and better understand the behavior of viruses.

These applications are perhaps too complex for a new math learner to master and require many years of math studies beyond algebra to understand them. However, learning algebra is the beginning of a journey that gives you the tools and the necessary skills to solve more complex systems. You might still ask why a person who is not planning to become a scientist, business analyst, or biomedical engineer must learn algebra. It is important because:

Algbera is a gateway course, used by other disciplines as a hurdle for college admission.

Algebra is part of our cultural legacy, much like art and music, and its mastery is expected by all who want to be considered educated.

Algebra strengthens deductive reasoning.

Algebra provides a first example of an *abstract system*.

Despite the above reasons, what if you still claim that you cannot learn algebra? Is algebra not for you? Unlike foreign languages, algebra is a more universal language, much like music, and to fully appreciate its power you must be willing to learn it. Most teachers will claim that anyone can learn algebra. Most mathematicians and mathematics instructors will state that the major difficulties in learning algebra stem from previous difficulties with arithmetic. Therefore, if you are reasonably versed in arithmetic, you should not have any trouble with learning algebra. Keep in mind that everyone is capable of learning algebra, so take it one step at a time and learn the basic rules before moving on to the more advanced concepts.

Supplementary Exercises

Exercise 1.1.1 *Comment on the following statement:* There is no need to learn mathematics, because today all calculations can be carried out by computers.

Exercise 1.1.2 *Comment on the following statement:* I was never good at math. I will never pass this algebra class.

Exercise 1.1.3 *Comment on the following statement:* Only nutritionists should know about the basics of nutrition, since that is their profession. Only medical doctors should know about the basics of health, since that is their profession. Only mathematicians should know about algebra, since that is their profession.

1.2 WHAT IS ELEMENTARY ALGEBRA ALL ABOUT?

Elementary algebra generalizes arithmetic by treating quantities in the abstract. In other words, mathematical expressions use variables (symbols) instead of values (numbers) to represent an algebraic equation. For example, consider two arithmetic problems: $2 + 3 = 3 + 2$ and $4 + 1 = 1 + 4$. In elementary algebra, our arithmetic problem can be expressed for any two numbers a, b and rewritten as

$$a + b = b + a.$$

Algebra allows the formulation of problems and their resolution by treating an unknown quantity formally with a variable or a symbol, such as *a* or *b*. For example, we will learn later how to calculate

$$123456789^2 - (123456787)(123456791)$$

without using a calculator. But the power of algebra goes beyond these curiosities.

Supplementary Exercises

Exercise 1.2.1 *You start with* **$100**. *You give* **20%** *to your friend. But it turns out that you need the* **$100** *after all in order to pay a debt. By what percent should you increase your current amount in order to restore the* **$100**? *(The answer is not* **20%**.)

Exercise 1.2.2 *A bottle of wine and its cork cost* **$1**. *The bottle of wine costs* 80¢ *more than the cork. What is the price of the cork, in cents?*

1.3 PUZZLES

The purpose of the puzzles in this section is to demonstrate some mathematical problem-solving techniques: working backwards, searching for patterns, and case-by-case analysis.

Example 1.1 A frog is in a **10** ft well. At the beginning of each day, it leaps **5** ft up, but at the end of the day it slides **4** ft down. After how many days, if at all, will the frog escape the well?

Solution: ▶ *The frog will escape after seven days. At the end of the sixth day, the frog has leaped* **6** *feet. Then at the beginning of the seventh day, the frog leaps* **5** *more feet and is out of the well.* ◀

Example 1.2 Dale should have divided a number by **4**, but instead he subtracted **4**. He got the answer **48**. What should his answer have been?

    Solution: ▶ *We work backwards. He obtained* **48** *from* **48 + 4 = 52**. *This means that he should have performed* **52 ÷ 4 = 13**. ◀

Example 1.3 When a number is multiplied by **3** and then increased by **16**, the result obtained is **37**. What is the original number?

    Solution: ▶ *We work backwards as follows. We obtained* **37** *by adding* **16** *to a number. Working backwards,* **37 − 16 = 21**. *We obtained this* **21** *by multiplying a number by* **3**. *Again working backwards, the number* **21 ÷ 3 = 7**. *Thus the original number was* **7**. ◀

Example 1.4 You and I play the following game. I tell you to write down three 2-digit integers between **10** and **89**. Then I write down three 2-digit integers of my choice. The answer comes to **297**, no matter which three integers you choose (my choice always depends on yours). For example, suppose you choose **12, 23, 48**. Then I choose **87, 76, 51**. You add

$$12 + 23 + 48 + 87 + 76 + 51 = 297.$$

Again, suppose you chose **33, 56, 89**. I then choose **66, 43, 10**. Observe that

$$33 + 56 + 89 + 66 + 43 + 10 = 297.$$

Explain how I choose my numbers so that the answer always comes up to be **297**?

    Solution: ▶ *Notice that I always choose my number so that when I add it to your number I get* **99**. *Therefore, I end up adding* **99** *three times — and* **3 × 99 = 297**. ◀

Example 1.5 What is the sum of all the positive integers from **1** to **100**? This problem can be written as the following expression:

$$1 + 2 + 3 + \cdots + 99 + 100.$$

    Solution: ▶ *Pair up the numbers into the fifty pairs, as follows:*

$$(100 + 1) = (99 + 2) = (98 + 3) = \cdots = (50 + 51).$$

This provides **50** *pairs that each add up to* **101**, *so the desired sum is* **101 × 50 = 5050**. *Another solution to this problem will be given in Example 6.23.* ◀

Supplementary Exercises

Exercise 1.3.1 *What could St. Augustine mean by mathematicians making prophecies? Could he have meant a profession other than mathematician?*

Exercise 1.3.2 *Can you find five even integers whose sum is* **25**?

Exercise 1.3.3 *Doris entered an elevator in a tall building. She went up* **4** *floors, down* **6** *floors, up* **8** *floors, and down* **10** *floors. She then found herself on the* **23**rd *floor. In what floor did she enter the elevator?*

Exercise 1.3.4 *A natural number is called a* palindrome *if it reads the same both forwards and backwards. For example,* **1221** *and* **100010001** *are palindromes. The palindrome* **10001** *is strictly between two other palindromes. Which two?*

Exercise 1.3.5 *In the figure below, each square represents a digit. Find the value of each missing digit.*

$$
\begin{array}{ccccc}
 & \blacksquare & 7 & 5 & \blacksquare & 6 \\
- & & \blacksquare & 5 & 6 & \blacksquare \\
\hline
 & 2 & 4 & \blacksquare & 7 & 5 \\
\end{array}
$$

Exercise 1.3.6 *Is it possible to replace the letter* a *in the square below so that every row has the same sum of every column?*

1	2	5
3	3	2
a	3	1

Exercise 1.3.7 *Fill each square in the figure below with exactly one number from*

{1, 2, 3, 4, 5, 6, 7, 8, 9}

so that the square becomes a magic square, that is, a square where every row has the same sum as every column, and as every diagonal.

Is there more than one solution?

Exercise 1.3.8 *A brother and a sister collected* **24** *coins. The brother collected twice as many coins as his sister. How many did each collect?*

ARITHMETIC REVIEW

CHAPTER 2

ARITHMETIC OPERATIONS

This chapter reviews the mathematical operations of addition, subtraction, multiplication, and division of numbers and introduces exponentiation and root extraction. It is expected that most of the material here will be familiar to you. Nevertheless, arithmetic operations are presented in this chapter in such a way so that algebraic generalization can be easily derived from them and applied into algebraic equations.

2.1 SYMBOLICAL EXPRESSION

The study of algebra begins by interpreting the meaning of its symbols. The use of symbols or *letters* denotes arbitrary numbers and frees us from the use of a long list of cases. For example, suppose that

$$1 + 0 = 1, \quad 2 + 0 = 2, \quad \frac{1}{2} + 0 = \frac{1}{2}, ...,$$

Since numbers are infinite, it is an impossible task to list all cases. Instead, the use of an abstract system to represent the statement above saves space. Instead of listing all the possible cases for infinite numbers, you simply replace a number with an arbitrary symbol such as x. Where x is a number, then

$$x + 0 = x,$$

without knowing the exact value of the arbitrary number x.

Normally the words *increase, increment,* and *augment,* are associated with addition. Thus, if x is an unknown number, the expression "a number increased by seven" is translated into symbols as $x + 7$. Later in this chapter, this expression can also be equivallenty writtened as $7 + x$.

Similarly, the words *decrease, decrement, diminish,* and *difference* are associated with subtraction. Thus if x is an unknown number, the expression "a number decreased by seven" is translated into symbols as $x - 7$. Note that this expression is different than "seven decreased by a certain number" or the expression $7 - x$, which is discussed later in this chapter.

Multiplication is commonly associated with the word *product*. Thus if x is an unknown number, the expression "the product of a certain number and seven" is translated into symbols as $7x$. Notice here the use of *juxtaposition* to denote the multiplication of a letter and a number. In other words, the use of the $\times$ (times) symbol or the $\cdot$ (central dot) symbol is omitted. The following are all equivalent:

$$7x, \ 7 \cdot x, \ 7(x), \ (7)(x), \ 7 \times x.$$

Notice that a reason for *not* using × when we use letters is to avoid confusion with the letter *x*. Also, the expression could have been written as *x*7, but this usage is uncommon.

☞ *Multiplication operation symbols are necessary in order to represent the product of two numbers. Thus, write the product as $5 \cdot 6 = 5 \times 6 = (5)(6) = 30$ so that the reader is not confused with the number 56.*

A few other words are used for multiplication by a specific factor. If the unknown quantity is *a*, then *twice* the unknown quantity is represented by **2a**. *Thrice, or to triple*, the unknown quantity is represented by **3a**. The *square* of an unknown quantity is that quantity multiplied by itself. For example, the square of *a* is *aa*, which is represented in short by a^2. Here *a* is the *base* and **2** is the *exponent*. The *cube* of an unknown quantity is that quantity multiplied by its square, so for example, the cube of *a* is *aaa*, which is represented in short by a^3. Here *a* is the *base* and **3** is the *exponent*.

The word *quotient* will generally be used to denote division. For example, the quotient of a number and **7** is denoted by *x* ÷ **7**, or equivalently by *x/7 or* $\frac{x}{7}$. The *reciprocal* of a number *x* is 1/x or $\frac{1}{x}$.

Example 2.1 If a number *x* is tripled and its value is increased by five, we obtain **3x + 5**.

Example 2.2 If *x* is larger than *y*, the difference between *x* and *y* is **x − y**. However, if *y* is larger than *x*, the difference between *x* and *y* is **y − x**.

Example 2.3 If *a* and *b* are two numbers, then their product is **ab**. Later we will see that this is the same as **ba**.

Example 2.4 The sum of the squares of *x* and *y* is $x^2 + y^2$. However, the square of the sum of *x* and *y* is $(x + y)^2$.

Example 2.5 The expression $2x - \frac{1}{x^2}$ can be translated as "twice a number is diminished by the reciprocal of its square."

Example 2.6 If *n* is an integer, its predecessor is **n − 1** and its successor is **n + 1**.

Example 2.7 You begin the day with **E** eggs. During the course of the day, you fry **Y** omelettes, each requiring **A** eggs. How many eggs are left?

Solution: ▶ *E − YA, since YA eggs are used in frying Y omelettes. Thus, E −YA eggs are left.* ◀

Example 2.8 An even natural number has the form **2a**, where *a* is a natural number. An odd natural number has the form **2a + 1**, where *a* is a natural number.

Example 2.9 A natural number divisible by **3** has the form **3a**, where *a* is a natural number. A natural number leaving remainder **1** upon division by **3** has the form **3a + 1**, where *a* is a natural number. A natural number leaving remainder **2** upon division by 3 has the form **3a + 2**, where *a* is a natural number.

Example 2.10 Find a formula for the *n*th term of the arithmetic progression

$$2, 7, 12, 17, \ldots.$$

Solution: ▶ *We start with 2, and then keep adding 5, thus*

$$2 = 2 + 5 \cdot 0$$

$$7 = 2 + 5 \cdot 1$$

$$12 = 2 + 5 \cdot 2$$

$$17 = 2 + 5 \cdot 3, \ldots.$$

The general term is therefore of the form $2 + 5(n-1)$, *where* $n = 1, 2, 3, \ldots$ *is a natural number.* ◀

Example 2.11 Find a formula for the nth term of the geometric progression

$$6, 12, 24, 48, \ldots.$$

Solution: ▶ *We start with* **6,** *and then keep multiplying by* **2,** *thus*

$$6 = 6 \cdot 2^0, \qquad 12 = 6 \cdot 2^1, \qquad 24 = 6 \cdot 2^2, \qquad 48 = 6 \cdot 2^3, \ldots.$$

The general term is therefore of the form $3 \cdot 2^{n-1}$, *where* $n = 1, 2, 3, \ldots$ *is a natural number.* ◀

Example 2.12 The law of formation and conjecture in a general formula can be represented as follows:

$$1 = 1,$$
$$1 + 2 = \frac{(2)(3)}{2},$$
$$1 + 2 + 3 = \frac{(3)(4)}{2},$$
$$1 + 2 + 3 + 4 = \frac{(4)(5)}{2},$$
$$1 + 2 + 3 + 4 + 5 = \frac{(5)(6)}{2}.$$

Solution: ▶ *Notice that the right-hand side consists of the last number on the left times its successor, which is then divided by* **2.** *Thus we are asserting that*

$$1 + 2 + 3 + \cdots + (n-1) + n = \frac{(n)(n+1)}{2}.$$

◀

Supplementary Exercises

Exercise 2.1.1 *If a person is currently* N *years old, what was his age* **20** *years ago?*

Exercise 2.1.2 *If a person is currently* N *years old, what will his age be in* **20** *years?*

Exercise 2.1.3 *You start with* x *dollars. Then you triple this amount, and finally you increase that resulting amount by* **10** *dollars. How many dollars do you now have?*

Exercise 2.1.4 *You start with* x *dollars. Then you add* **10** *dollars to this amount, and finally you triple the resulting amount. How many dollars do you now have?*

Exercise 2.1.5 *A crocheted baby blanket uses* **5** *balls of yarn. I start the day with* b *balls of yarn and crochet* s *blankets. How many balls of yarn do I have at the end of the day?*

Exercise 2.1.6 *Think of a number and then double it. Next, add* **10.** *Half your result, and then subtract your original number. After these five steps, your answer is* **5** *regardless of your original number! If* x *is the original number, explain by means of algebraic formula each step.*

Exercise 2.1.7 *What is the general form for a natural number divisible by* **4?** *Leaving remainder* **1** *upon division by* **4?** *Leaving remainder* **2** *upon division by* **4?** *Leaving remainder* **3** *upon division by* **4?**

Exercise 2.1.8 *Find a general formula for the* nth *term of the arithmetic progression*

$$1, 7, 13, 19, 25, \ldots.$$

Exercise 2.1.9 *Identify the law of formation and conjecture a general formula:*

$$1^2 = \frac{(1)(2)(3)}{6},$$

$$1^2 + 2^2 = \frac{(2)(3)(5)}{6},$$

$$1^2 + 2^2 + 3^2 = \frac{(3)(4)(7)}{6},$$

$$1^2 + 2^2 + 3^2 + 4^2 = \frac{(4)(5)(9)}{6},$$

$$1^2 + 2^2 + 3^2 + 4^2 + 5^2 = \frac{(5)(6)(11)}{6}.$$

Exercise 2.1.10 *Identify the law of formation and conjecture a general formula.*

$$1 = 1^2,$$

$$1 + 3 = 2^2,$$

$$1 + 3 + 5 = 3^2,$$

$$1 + 3 + 5 + 7 = 4^2,$$

$$1 + 3 + 5 + 7 + 9 = 5^2.$$

Exercise 2.1.11 *You start the day with q quarters and d dimes. How much money do you have? Answer in cents. If by the end of the day you have lost a quarters and b dimes, how much money do you now have? Answer in cents.*

Exercise 2.1.12 *Let x be an unknown quantity. Translate into symbols the expression "the cube of a quantity is reduced by its square, and then what is left is divided by 8."*

Exercise 2.1.13 *A man bought a hat for h dollars. He then bought a jacket and a pair of jeans. If the jacket is thrice as expensive as the hat, and the jeans are 8 dollars cheaper than the jacket, how much money did he spend in total for the three items?*

Exercise 2.1.14 *A farmer sold a cow for a dollars and gained b dollars as profit. What is the real cost of the cow?*

Exercise 2.1.15 *A salesman started the year with m dollars. The first month he gained x dollars, the next month he lost y dollars, the third month he gained b dollars, and the fourth month lost z dollars. How much did he have at the end of the fourth month?*

2.2 THE NATURAL NUMBERS

This section gives an overview of the natural numbers. The natural numbers are typically used to count and to order objects. Start with two symbols, **0** and **1**, and an operation +, adjoining the elements

$$1 + 1,$$

$$1 + 1 + 1,$$

$$1 + 1 + 1 + 1,$$

$$1 + 1 + 1 + 1 + 1, \dots.$$

Observe that this set is *infinite* and *ordered*, that is, you can compare any two elements and tell whether one is larger than the other. Define the symbols

$$2 = 1 + 1,$$

$$3 = 1 + 1 + 1,$$

$$4 = 1 + 1 + 1 + 1,$$

$$5 = 1 + 1 + 1 + 1 + 1,$$

$$6 = 1 + 1 + 1 + 1 + 1 + 1,$$

$$7 = 1 + 1 + 1 + 1 + 1 + 1 + 1,$$

$$8 = 1 + 1 + 1 + 1 + 1 + 1 + 1 + 1,$$

$$9 = 1 + 1 + 1 + 1 + 1 + 1 + 1 + 1 + 1.$$

Beyond **9,** reuse these symbols by also attaching a meaning to their place. Thus

$$10 = 1 + 1 + 1 + 1 + 1 + 1 + 1 + 1 + 1 + 1,$$

$$11 = 1 + 1 + 1 + 1 + 1 + 1 + 1 + 1 + 1 + 1 + 1,$$

$$12 = 1 + 1 + 1 + 1 + 1 + 1 + 1 + 1 + 1 + 1 + 1 + 1,$$

Definition 2.1 A *positional notation* or *place-value notation system* is a numeral system in which each position is related to the next by a constant multiplier of that numeral system. Each position is represented by a limited set of symbols. The resultant value of each position is the value of its symbol or symbols multiplied by a power of the base.

As you know, we use base-10 positional notation. For example, in

$$1234 = 1 \cdot 1000 + 2 \cdot 100 + 3 \cdot 10 + 4 \cdot 1,$$

1 does not mean "**1**," but **1000**; **2** does not mean "**2**," but **200**, etc.

Before *positional notation* became standard, simple additive systems (sign-value notation) were used, such as the value of the Hebrew letters, the value of the Greek letters, and Roman Numerals. Arithmetic with these systems was incredibly cumbersome.

Definition 2.2 The collection of all numbers defined by the recursion method above is called the set of *natural numbers,* and we represent them by the symbol N, that is,

$$\mathbb{N} = \{0, 1, 2, 3, ...\}.$$

Natural numbers are used for two main reasons:

1. Counting, for example, "There are **10** sheep in the herd."

2. Ordering, for example, "Los Angeles is the second largest city in the United States."

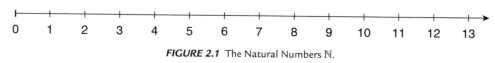

FIGURE 2.1 The Natural Numbers $\mathbb{N}$.

The natural numbers are interpreted as a linearly ordered set of points, as shown in Figure 2.1, to induce an order relation as defined below.

Definition 2.3 Let **a** and **b** be two natural numbers. We say that **a** *is (strictly) less than* **b,** if **a** is to the left of **b** on the natural number line. We denote this by **a** < **b**.

☞ *The symbol* ∈ *is used to indicate that a certain element belongs to a certain set. The negation of* ∈ *is* ∉. *For example,* **1** ∈ ℕ *because* **1** *is a natural number, but* $\frac{1}{2}$ ∉ ℕ *because* $\frac{1}{2}$ *is not a natural number.*

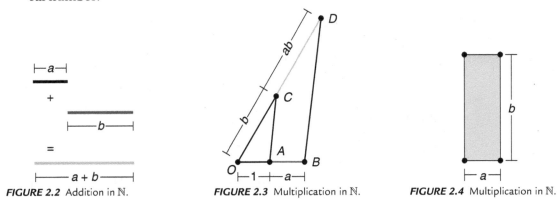

FIGURE 2.2 Addition in ℕ.   FIGURE 2.3 Multiplication in ℕ.   FIGURE 2.4 Multiplication in ℕ.

You may think of the addition of natural numbers as concatenation of segment lengths. For example, in Figure 2.2 a segment whose length is **b** units is added to a segment whose length is **a** units, resulting in a new segment whose length is **a + b** units.

Multiplication is somewhat harder to interpret. In Figure 2.3, consider form △**OAC** with segment **OA = 1** and segment **OC = b**. Extend the segment **[OA]** to point **B**, with **AB = a**. Through point **B** draw a line parallel to segment **[AC]**, meeting segment **[OC]** extended at point **D**. By the similarity of △**OAC** and △**OBD**, segment **CD = ab**. Another interpretation of multiplication occurs in Figure 2.4, where a rectangle of area **ab** (square units) is formed having sides of **a** units by **b** units.

Observe that if we add or multiply any two natural numbers, the result is a natural number. We encode this observation in the following axiom.

Axiom 2.1 (Closure) ℕ is *closed under addition,* that is, if **a** ∈ ℕ and **b** ∈ ℕ then also **a + b** ∈ ℕ. ℕ is *closed under multiplication,* that is, if **x** ∈ ℕ and **y** ∈ ℕ then also **x y** ∈ ℕ.

If **0** is added to any natural number, the result is unchanged. Similarly, if a natural number is multiplied by **1**, the result is also unchanged. This is encoded in the following axiom.

Axiom 2.2 (Additive and Multiplicative Identity) **0** ∈ ℕ is the *additive identity* of ℕ, that is, it has the property that for all **x** ∈ ℕ it follows that

$$x = 0 + x = x + 0.$$

1 ∈ ℕ is the *multiplicative identity* of ℕ, that is, it has the property that for all **a** ∈ ℕ it follows that

$$a = 1a = a1.$$

Again, it is easy to see that when two natural numbers are added or multiplied, the result does not depend on the order. This is encoded in the following axiom.

Axiom 2.3 (Commutativity) Let **a** ∈ ℕ and **b** ∈ ℕ. Then **a + b = b + a** and **ab = ba**.

Two other important axioms for the natural numbers are now given.

Axiom 2.4 (Associativity) Let **a**, **b**, **c** be natural numbers. Then the order of parentheses when performing addition is irrelevant, that is,

$$a + (b + c) = (a + b) + c = a + b + c.$$

Similarly, the order of parentheses when performing multiplication is irrelevant,

$$a(bc) = (ab)c = abc.$$

Axiom 2.5 (Distributive Law) Let a, b, c be natural numbers. Then

$$a(b + c) = ab + ac,$$

and

$$(a + b)c = ac + bc.$$

The product mn is, simply put, shorthand for addition. That multiplication has the equivalent expressions

$$mn = \underbrace{n + n + \cdots + n}_{m \text{ times}} = \underbrace{m + m + \cdots + m}_{n \text{ times}}.$$

Thus $(3)(4)$ means the following expression

$$(3)(4) = 3 + 3 + 3 + 3 = 4 + 4 + 4 = 12.$$

Therefore solving the expression

$$(3)(5) + (6)(4)$$

requires performing the multiplication first and then the addition, obtaining

$$(3)(5) + (6)(4) = 5 + 5 + 5 + 6 + 6 + 6 + 6 = 39,$$

or more concisely

$$(3)(5) + (6)(4) = 15 + 24 = 39.$$

In turn, a shorter form for multiplication by the same number is exponentiation, which is defined as follows:

Definition 2.4 (Exponentiation) If n is a natural number greater than or equal to **1**, then the *nth power of a* is defined by

$$a^n = \underbrace{a \cdot a \cdots a}_{n \text{ times}}$$

Here, a is the *base* and n is the *exponent*. If a is any number different from **0**, then we define

$$a^0 = 1.$$

without attaching any meaning to 0^0.

Example 2.13 Powers of **2** is a very important concept in computers. A *bit* is a binary digit taking a value of either **0** (electricity does not pass through a circuit) or **1** (electricity passes through a circuit). Then,

$$
\begin{aligned}
2^1 &= 2 & 2^6 &= 64 \\
2^2 &= 4, & 2^7 &= 128 \\
2^3 &= 8, & 2^8 &= 256 \\
2^4 &= 16, & 2^9 &= 512 \\
2^5 &= 32, & 2^{10} &= 1024.
\end{aligned}
$$

Since $2^{10} \approx 1000$, 2^{10} is referred to as a *kilobit*.

Example 2.14 Notice that $2^3 = 8$ and $3^2 = 9$ are consecutive powers.

☞ *Notice that a^b is not ab. Thus $2^3 = (2)(2)(2) = 8$, and not $(2)(3) = 6$.*

In any expression containing addition and exponentiation, the exponentiation is performed first, since it is really a shortcut for writing multiplication.

Example 2.15

$$(2)(4) + 3^3 = 8 + 27 = 35,$$

$$3^2 + 2^3 = 9 + 8 = 17,$$

$$(3^2)(4)(5) = (9)(4)(5) = 180.$$

The order of operations can be forced by grouping symbols, like parentheses ( ), brackets [ ], or braces { }.

Example 2.16

$$(3 + 2)(5 + 3) = (5)(8) = 40,$$

$$(3 + 2)^2 = (5)^2 = 25,$$

$$(5 + (3 + 2(4))^2)^3 = (5 + (3 + 8)^2)^3 = (5 + (11)^2)^3 = (5 + 121)^3 = 126^3 = 2000376.$$

☞ *Observe that $(3 + 2)^2 = 25$ but that $3^2 + 2^2 = 9 + 4 = 13$. Thus exponentiation does not distribute over addition.*

Example 2.17 Each element of the set

$$\{10, 11, 12, ..., 19, 20\}$$

is multiplied by each element of the set

$$\{21, 22, 23, ..., 29, 30\}.$$

If all these products are added, what is the resulting sum?

Solution: ▶ *This is asking for the product $(10 + 11 + \cdots + 20)(21 + 22 + \cdots + 30)$ after all the terms are multiplied. But $10 + 11 + \cdots + 20 = 165$ and $21 + 22 + \cdots + 30 = 255$. Therefore we want $(165)(255) = 42{,}075$.* ◀

Definition 2.5 To *evaluate* an expression with letters means to substitute the values of its letters by the equivalent values given.

Example 2.18 Evaluate $a^3 + b^3 + c^3 + 3abc$ when $a = 1, b = 2, c = 3$.

Solution: ▶ *Substituting the variables a, b, c with their given values, 1, 2, and 3,*

$$1^3 + 2^3 + 3^3 + 3(1)(2)(3) = 1 + 8 + 27 + 18 = 54.$$ ◀

The root is the reverse mathematical operation of exponentiation and can be defined as follows:

Definition 2.6 (Roots) Let m be a natural number greater than or equal to 2, and let a and b be any natural numbers. We write that $\sqrt[m]{a} = b$ if $a = b^m$. In this case we say that b is the *mth root* of a. The number m is called the *index* of the root.

☞ *In the special case when $m = 2$, we do not write the index. Thus we will write $\sqrt{a}$ rather than $\sqrt[2]{a}$. The number $\sqrt{a}$ is called the square root of a. The number $\sqrt[3]{a}$ is called the cubic root of a.*

Example 2.19 Let

$$\sqrt{1} = 1 \quad \text{because} \quad 1^2 = 1,$$
$$\sqrt{4} = 2 \quad \text{because} \quad 2^2 = 4,$$
$$\sqrt{9} = 3 \quad \text{because} \quad 3^2 = 9,$$
$$\sqrt{16} = 4 \quad \text{because} \quad 4^2 = 16,$$
$$\sqrt{25} = 5 \quad \text{because} \quad 5^2 = 25,$$
$$\sqrt{36} = 6 \quad \text{because} \quad 6^2 = 36.$$

Example 2.20 Let

$$\sqrt[10]{1} = 1 \quad \text{because} \quad 1^{10} = 1,$$
$$\sqrt[5]{32} = 2 \quad \text{because} \quad 2^5 = 32,$$
$$\sqrt[3]{27} = 3 \quad \text{because} \quad 3^3 = 27,$$
$$\sqrt[3]{64} = 4 \quad \text{because} \quad 4^3 = 64,$$
$$\sqrt[3]{125} = 5 \quad \text{because} \quad 5^3 = 125.$$
$$\sqrt[10]{1024} = 2 \quad \text{because} \quad 2^{10} = 1024.$$

Having an idea of what it means to add and multiply natural numbers, let us define subtraction and division of natural numbers by means of those operations. This is often the case in mathematics: we define a new procedure in terms of old procedures.

Definition 2.7 (Definition of Subtraction) Let m, n, x be natural numbers. Then the statement $m - n = x$ means that $m = x + n$.

Example 2.21 To compute $15 - 3$ think of which number when added to 3 returns 15, clearly then $15 - 3 = 12$ since $15 = 12 + 3$.

Definition 2.8 (Definition of Division) Let m, n, x be natural numbers, with $n \neq 0$. Then the statement $m \div n = x$ means that $m = xn$.

Example 2.22 To compute $15 \div 3$, think of which number when multiplied by 3 returns 15. Clearly then $15 \div 3 = 5$ since $15 = 5 \cdot 3$.

☞ *Neither subtraction nor division are included in* $\mathbb{N}$. *For example,* $3 - 5$ *is not a natural number, and neither is* $3 \div 5$. *Again, the operations of subtraction and division misbehave in the natural numbers; they are not commutative. For example,* $5 - 3$ *is not the same as* $3 - 5$ *and* $20 \div 4$ *is not the same as* $4 \div 20$.

Supplementary Exercises

Exercise 2.2.1 *Find the numerical value of* $1^1 2^2 3^3$.

Exercise 2.2.2 *Find the numerical value of* $(\sqrt{36} - \sqrt{25})^2$.

Exercise 2.2.3 *Find the numerical value of* $3 \cdot 4 + 4^2$.

Exercise 2.2.4 *Evaluate* $(a + b)(a - b)$ *when* $a = 5$ *and* $b = 2$.

Exercise 2.2.5 *Evaluate* $(a^2 + b^2)(a^2 - b^2)$ *when* $a = 2$ *and* $b = 1$.

Exercise 2.2.6 *If today is Thursday, what day will it be* **100** *days from now?*

Exercise 2.2.7 *If the figure shown below is folded to form a cube, then three faces meet at every vertex. If for each vertex we take the product of the numbers on the three faces that meet there, what is the largest product we get?*

Exercise 2.2.8 *A freelance graphic designer must print* **4500** *cards. One contractor can print the cards in* **30** *days and another contractor can print them in* **45** *days. How many days would be needed if both contractors are working simultaneously?*

Exercise 2.2.9 *For a collegiate commencement ceremony, the students of a school are arranged in* **9** *rows of* **28** *students per row. How many rows could be made with* **36** *students per row?*

Exercise 2.2.10 *Oscar rides his bike, being able to cover* **6** *miles in* **54** *minutes. At that speed, how long does it take him to travel a mile?*

Exercise 2.2.11 *For which values of the natural number* **n** *is* **36 ÷ n** *a natural number?*

Exercise 2.2.12 *Doing only* one *multiplication, prove that*

$$(666)(222) + (1)(333) + (333)(222) + (666)(333) + (1)(445) + (333)(333)$$
$$+ (666)(445) + (333)(445) + (1)(222) = 1000000.$$

Exercise 2.2.13 *A truck with* **5** *tires (four road tires and a spare tire) travelled* **30,000** *miles. If all five tires were used equally, how many miles of wear did each tire receive?*

Exercise 2.2.14 *A quiz has* **25** *questions with* **4** *points awarded for each correct answer and* **1** *point deducted for each incorrect answer, with zero points for each question omitted. Anna scores* **77** *points. How many questions did she omit?*

Exercise 2.2.15 *A calculator displays the computed result of the product*

$$987654 \times 745321$$

as the number **7.36119E11,** *which means* **736,119,000,000.** *Explain the process to find the last six missing digits.*

Exercise 2.2.16 *How many digits does* $4^{16}5^{25}$ *have?*

Exercise 2.2.17 *A farmer has decided to pose the following problem: "If one gathers all of my cows into groups of* **4, 5,** *or* **6,** *there will be no remainder. But if one gathers them into groups of* **7** *cows, there will be* **1** *cow left in one group." The number of cows is the smallest positive integer satisfying these properties. How many cows are there?*

Exercise 2.2.18 *Create a new arithmetic operation* $\oplus$ *by letting* $a \oplus b = 1 + ab.$

   1. *Compute* $1 \oplus (2 \oplus 3).$

   2. *Compute* $(1 \oplus 2) \oplus 3.$

   3. *Is your operation associative? Explain.*

   4. *Is the operation commutative? Explain.*

2.3 FRACTIONS

This section provides a review of fractions in terms of arithmetic operations.

Definition 2.9 A *(positive numerical) fraction* is a number of the form $m \div n = \frac{m}{n}$ where m and n are natural numbers and $n \neq 0$. Here m is the *numerator* of the fraction and n is the *denominator* of the fraction.

Given a natural number $n \neq 0$, we divide the interval between consecutive natural numbers k and $k+1$ into n equal pieces. Figures 2.5, 2.6, and 2.7 show examples with $n = 2$, $n = 3$, and $n = 4$, respectively. Notice that the larger n is, the finer the partition.

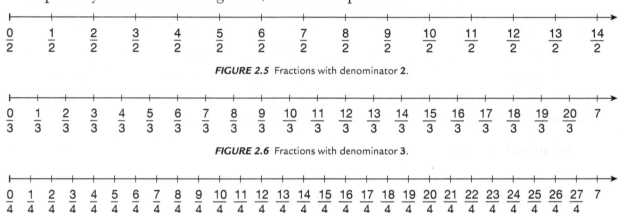

FIGURE 2.5 Fractions with denominator 2.

FIGURE 2.6 Fractions with denominator 3.

FIGURE 2.7 Fractions with denominator 4.

☞ *Notice that in many of the fractions written above, the numerator is larger than the denominator, commonly called* improper fractions. *In many instances, improper fractions are converted into to mixed numbers. For example,* $\frac{9}{2}$ *is an improper fraction that can also be rewritted into a mixed number as* $4\frac{1}{2}$ *. In this book, improper fractions are preferred since it is the common usage in Algebra and in modern computer algebra programs.*

By doing more of the diagrams above you may have noticed that there are multiple names for, say, the natural number **2**. For example

$$2 = \frac{2}{1} = \frac{4}{2} = \frac{6}{3} = \frac{8}{4} = \frac{10}{5} = \frac{12}{6},$$

This observation is a particular case of the following theorem.

Theorem 2.1 (Cancellation Law) Let m, n, k be natural numbers with $n \neq 0$ and $k \neq 0$. Then

$$\frac{mk}{nk} = \frac{m}{n}.$$

Thus given a fraction, if the numerator and the denominator have any common factors greater than **1**, that is, any nontrivial factors, we may reduce the fraction and get an equal fraction.

Definition 2.10 Two fractions such that $\frac{a}{b} = \frac{x}{y}$ are said to be *equivalent*. If $b < y$, then $\frac{a}{b}$ is said to be a *reduced form* of $\frac{x}{y}$.

☞ *It is possible to prove that any fraction has a unique reduced form with minimal denominator, the so-called* equivalent fraction in lowest terms. *This depends on the fact that the natural numbers can be factored uniquely into primes. Although we will accept this result, we will not prove it here.*

Example 2.23 To reduce $\frac{104}{120}$ to lowest terms, observe that

$$\frac{104}{120} = \frac{104 \div 4}{120 \div 4} = \frac{26 \div 2}{30 \div 2} = \frac{13}{15}.$$

Since **13** and **15** do not have a nontrivial factor in common, $\frac{13}{15}$ is the desired reduction.

☞ *The reduction steps above are not unique. For example, if we had seen right away that **8** was a common factor, we would have obtained* $\frac{104}{120} = \frac{104 \div 8}{120 \div 8} = \frac{13}{15}$, *obtaining the same result. No matter how many steps you take, as long as there is valid cancellation to do and you perform them all, you will always obtain the correct result.*

Example 2.24 Find a fraction with denominator **120** equivalent to $\frac{11}{24}$.

Solution: ▶ *Observe that* **120 ÷ 24 = 5**. *Thus*

$$\frac{11}{24} = \frac{11 \cdot 5}{24 \cdot 5} = \frac{55}{120}.$$

◀

Supplementary Exercises

Exercise 2.3.1 *Four comrades are racing side by side down a dusty staircase. Freddy goes down two steps at a time, Lorance three, George and Aragorn five. If the only steps with the footprints of all four comrades are at the top and the bottom, how many steps have just one footprint?*

Exercise 2.3.2 *What fraction of an hour is **36** minutes?*

Exercise 2.3.3 *Express $\frac{102}{210}$ in least terms.*

Exercise 2.3.4 *Find an equivalent fraction to $\frac{102}{210}$ with denominator **3990**.*

Exercise 2.3.5 *Arrange in increasing order: $\frac{2}{3}, \frac{3}{5}, \frac{8}{13}$.*

Exercise 2.3.6 *Four singers take part in a musical round of **4** equal lines, each finishing after singing the round through **3** times. The second singer begins when the first singer begins the second line, the third singer begins when the first singer begins the third line, the fourth singer begins when the first singer begins the fourth line. What is the fraction of the total singing time when all the singers are singing simultaneously?*

2.4 OPERATIONS WITH FRACTIONS

We now define addition of fractions. We would like this definition to agree with our definition of addition of natural numbers. Recall that we defined addition of natural numbers x and y as the concatenation of two segments of length x and y. We then define the addition of two fractions $\frac{a}{b}$ and $\frac{c}{d}$ as the concatenation of two segments of length $\frac{a}{b}$ and $\frac{c}{d}$. The properties of associativity and commutativity stem from this definition. The problem we now have is how to concretely apply this definition to find the desired sum. From this concatenation definition it follows that natural numbers a and $b \neq 0$.

$$\frac{a}{b} = \underbrace{\frac{1}{b} + \cdots + \frac{1}{b}}_{a\ \text{times}}.$$

It follows from the above property that

$$\frac{x}{b} + \frac{y}{b} = \frac{x+y}{b}.$$

We now determine a general formula for adding fractions of different denominators.

Theorem 2.2 (Sum of Fractions) Let a,b,c,d be natural numbers with $b \neq 0$ and $d \neq 0$. Then

$$\frac{a}{b} + \frac{c}{d} = \frac{ad + bc}{bd}.$$

Observe that the trick for adding the fractions in the preceding theorem was to convert them to fractions of the same denominator.

Definition 2.11 To express two fractions in a *common denominator* is to write them in the same denominator. The smallest possible common denominator is called the *least common denominator*.

Example 2.25 Add $\frac{3}{5} + \frac{4}{7}$.

Solution: ▶ *A common denominator is* $5 \cdot 7 = 35$. *We thus find*

$$\frac{3}{5} + \frac{4}{7} = \frac{3 \cdot 7}{5 \cdot 7} + \frac{4 \cdot 5}{7 \cdot 5} = \frac{21}{35} + \frac{20}{35} = \frac{41}{35}. \qquad ◀$$

☞ *In the preceding example,* **35** *is not the only denominator that we may have used. Observe that* $\frac{3}{5} = \frac{42}{70}$ *and* $\frac{4}{7} = \frac{40}{70}$. *Adding,*

$$\frac{3}{5} + \frac{4}{7} = \frac{3 \cdot 14}{5 \cdot 14} + \frac{4 \cdot 10}{7 \cdot 10} = \frac{42}{70} + \frac{40}{70} = \frac{82}{70} = \frac{82 \div 2}{70 \div 2} = \frac{41}{35}.$$

This shows that it is not necessary to find the least common denominator *in order to add fractions, simply a common denominator.*

In fact, let us list the multiples of 5 and of 7 and let us underline the common multiples on these lists:

The multiples of 5 are   **5, 10, 15, 20, 25, 30, 35, 40, 45, 50, 55, 60, 65, 70, 75, ...,**

The multiples of 7 are   **7, 14, 21, 28, 35, 42, 49, 56, 63, 70, 77, ...**

The sequence

$$\textbf{35, 70, 105, 140, ...,}$$

is a sequence of common denominators for **5** and **7**.

Example 2.26 To perform the addition $\frac{2}{7}+\frac{1}{5}+\frac{3}{2}$, observe that $7\cdot 5\cdot 2=70$ is a common denominator. Thus

$$\frac{2}{7}+\frac{1}{5}+\frac{3}{2}=\frac{2\cdot 10}{7\cdot 10}+\frac{1\cdot 14}{5\cdot 14}+\frac{3\cdot 35}{2\cdot 35}$$

$$=\frac{20}{70}+\frac{14}{70}+\frac{105}{70}$$

$$=\frac{20+14+105}{70}$$

$$=\frac{139}{70}.$$

A few words about subtraction of fractions. Suppose $\frac{a}{b}\geq\frac{c}{d}$. Observe that this means that $ad\geq bc$, which in turns means that $ad-bc\geq 0$. Then from a segment of length $\frac{a}{b}$ we subtract one of $\frac{b}{d}$. In much the same manner of addition of fractions, then

$$\frac{a}{b}-\frac{c}{d}=\frac{ad}{bd}-\frac{bc}{bd}=\frac{ad-bc}{bd}.$$

Now, since $ad-bc\geq 0$, $ad-bc$ is a natural number and so $\frac{ad-bc}{bd}$ is a fraction.

We now approach multiplication of fractions. We saw that a possible interpretation for the product xy of two natural numbers x and y is the area of a rectangle with sides of length x and y. We would like to extend this interpretation in the case when x and y are fractions. That is, given a rectangle of sides of length $x=\frac{a}{b}$ and $y=\frac{c}{d}$, we would like to deduce that $\frac{ac}{bd}$ is the area of this rectangle.

Theorem 2.3 (Multiplication of Fractions) Let a, b, c, d be natural numbers with $b\neq 0$ and $d\neq 0$. Then

$$\frac{a}{b}\cdot\frac{c}{d}=\frac{ac}{bd}.$$

Example 2.27 We have

$$\frac{2}{3}\cdot\frac{3}{7}=\frac{6}{21}=\frac{2}{7}.$$

Alternatively, we could have cancelled the common factors, as follows,

$$\frac{2}{\cancel{3}}\cdot\frac{\cancel{3}}{7}=\frac{2}{7}.$$

Example 2.28 Find the exact value of the product

$$\left(1-\frac{2}{5}\right)\left(1-\frac{2}{7}\right)\left(1-\frac{2}{9}\right)\cdots\left(1-\frac{2}{99}\right)\left(1-\frac{2}{101}\right).$$

Solution: ▶ *We have,*

$$\left(1-\frac{2}{5}\right)\left(1-\frac{2}{7}\right)\left(1-\frac{2}{9}\right)\cdots\left(1-\frac{2}{99}\right)\left(1-\frac{2}{101}\right)$$

$$=\left(\frac{5}{5}-\frac{2}{5}\right)\left(\frac{7}{7}-\frac{2}{7}\right)\left(\frac{9}{9}-\frac{2}{9}\right)\cdots\left(\frac{99}{99}-\frac{2}{99}\right)\left(\frac{101}{101}-\frac{2}{101}\right)$$

$$=\frac{3}{5}\cdot\frac{5}{7}\cdot\frac{7}{9}\cdot\frac{9}{11}\cdots\frac{97}{99}\cdot\frac{99}{101}$$

$$=\frac{3}{101}.$$

◀

We now tackle division of fractions. Recall that we defined division of natural numbers as follows. If $n \neq 0$ and m, x are natural numbers, then $m \div n = x$ means that $m = xn$. We would like a definition of fraction division compatible with this definition of natural number division. Therefore, we give the following definition.

Definition 2.12 Let a, b, c, d be natural numbers with $b \neq 0$, $c \neq 0$, $d \neq 0$. We define the fraction division

$$\frac{a}{b} \div \frac{c}{d} = \frac{x}{y} \iff \frac{a}{b} = \frac{x}{y} \cdot \frac{c}{d}.$$

We would like to know what $\frac{x}{y}$ above is in terms of a, b, c, d. For this purpose we have the following theorem.

Theorem 2.4 (Division of Fractions) Let a, b, c, d be natural numbers with $b \neq 0$, $c \neq 0$, $d \neq 0$. Then

$$\frac{a}{b} \div \frac{c}{d} = \frac{a}{b} \cdot \frac{d}{c} = \frac{ad}{bc},$$

that is, $\frac{x}{y}$ in Definition 2.12 is $\frac{x}{y} = \frac{ad}{bc}$.

Definition 2.13 Let $c \neq 0$, $d \neq 0$ be natural numbers. The *reciprocal* of the fraction $\frac{c}{d}$ is the fraction $\frac{c}{d}$.

Theorem 2.4 says that in order to divide two fractions we must simply multiply the first one by the reciprocal of the other.

Example 2.29 We have,

$$\frac{24}{35} \div \frac{20}{7} = \frac{24}{35} \cdot \frac{7}{20} = \frac{4 \cdot 6}{7 \cdot 5} \cdot \frac{7 \cdot 1}{4 \cdot 5} = \frac{6 \cdot 1}{5 \cdot 5} = \frac{6}{25}.$$

Supplementary Exercises

Exercise 2.4.1 *Complete the "fraction puzzle" below.*

$\frac{1}{3}$	+		=	2
+	■	−	■	
$\frac{3}{4}$	×		=	
=	■	=	■	
	÷	$\frac{2}{3}$	=	

Exercise 2.4.2 *Find the exact numerical value of*

$$\left(\frac{2}{51}\right) \div \left(\frac{3}{17}\right)\left(\frac{7}{10}\right).$$

Exercise 2.4.3 *Find the exact numerical value of*

$$\frac{\dfrac{4}{7} - \dfrac{2}{5}}{\dfrac{4}{7} + \dfrac{2}{5}}.$$

Exercise 2.4.4 *Find the value of*

$$\frac{10 + 10^2}{\dfrac{1}{10} + \dfrac{1}{100}}.$$

Exercise 2.4.5 *Find the exact numerical value of*

$$\frac{1^3 + 2^3 + 3^3 - 3(1)(2)(3)}{(1+2+3)(1^2 + 2^2 + 3^2 - 1 \cdot 2 - 2 \cdot 3 - 3 \cdot 1)}.$$

Exercise 2.4.6 *If*

$$\frac{1}{1 + \dfrac{1}{5}} = \frac{a}{b},$$

where the fraction $\frac{a}{b}$ is in least terms, find $a^2 + b^2$.

Exercise 2.4.7 *Find the exact numerical value of*

$$\left(1 - \frac{1}{2}\right)\left(1 - \frac{1}{3}\right)\left(1 - \frac{1}{4}\right)\cdots\left(1 - \frac{1}{99}\right)\left(1 - \frac{1}{100}\right).$$

Exercise 2.4.8 *John takes* **2** *hours to paint a room, whereas Bill takes* **3** *hours to paint the same room. How long would it take if both of them start and work simultaneously?*

Exercise 2.4.9 *Naomi has* **16** *yards of gift-wrap in order to wrap the gifts for the Festival of Lights at the community center. Each gift requires* $1\frac{7}{8}$ *yards of paper. How many gifts can she wrap?*

Exercise 2.4.10 *Evaluate*

$$\cfrac{1}{2-\cfrac{1}{2-\cfrac{1}{2-\cfrac{1}{2}}}}.$$

Exercise 2.4.11 *Find the value of*

$$1+\cfrac{1}{2+\cfrac{1}{3+\cfrac{1}{4}}}.$$

Exercise 2.4.12 *At a college* **99%** *of the* **100** *students are female, but only* **98%** *of the students living on campus are female. If some females live on campus, how many students live off campus?*

Exercise 2.4.13 *What would be the price of a* $5\frac{1}{2}$*-mile trip with the following taxi company?*

SMILING CAMEL TAXI SERVICES	
First $\frac{1}{4}$ mi	$ .85
Additional $\frac{1}{4}$ mi	$ .40

2.5 THE INTEGERS

The introduction of fractions in the preceding section helped solve the problem that the natural numbers are not closed under division. We now solve the problem that the natural numbers are not closed under subtraction.

Definition 2.14 A natural number not equal to **0** is said to be *positive*. The set

$$\{1, 2, 3, 4, 5, ...\}$$

is called the set of *positive integers*.

Definition 2.15 Given a natural number **n**, we define its opposite **−n** as the unique number **−n** such that

$$n + (-n) = (-n) + n = 0.$$

The collection

$$\{-1, -2, -3, -4, -5, ...\}$$

of all the opposites of the natural numbers is called the set of *negative integers*. The collection of natural numbers together with the negative integers is the set of *integers*, which we denote by the symbol $\mathbb{Z}$. A graphical representation of the integers is given in Figure 2.8.

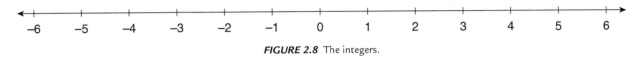

$$-6 \quad -5 \quad -4 \quad -3 \quad -2 \quad -1 \quad 0 \quad 1 \quad 2 \quad 3 \quad 4 \quad 5 \quad 6$$

FIGURE 2.8 The integers.

There seems to be no evidence of usage of negative numbers by the Babylonians, Pharaonic Egyptians, or the ancient Greeks. It seems that the earliest usage of them came from China and India. In the seventh century, negative numbers were used for bookkeeping in India. The Hindu astronomer Brahmagupta, writing around a.d. 630, shows a clear understanding of the usage of negative numbers.

Thus it took humans a few millennia to develop the idea of negative numbers. Perhaps because our lives are more complex now, it is not so difficult for us to accept their existence and understand the concept of negative numbers.

Let $a \in \mathbb{Z}$ and $b \in \mathbb{Z}$. If $a > 0$, then $-a < 0$. If $b < 0$, then $-b > 0$. Thus either the number, or its "mirror reflection" about 0, is positive, and in particular, for any $a \in \mathbb{Z}$, $-(-a) = a$. This leads to the following definition.

Definition 2.16 Let $a \in \mathbb{Z}$. The *absolute value* of a is defined and denoted by

$$|a| \begin{cases} a & \text{if } a \geq 0, \\ -a & \text{if } a < 0, \end{cases}$$

Example 2.30 $|5| = 5$ since $5 > 0$. $|-5| = -(-5) = 5$, since $-5 < 0$.

☞ *Letters have no idea of the sign of the numbers they represent. Thus it is a* **mistake** *to think, say, that* $+x$ *is always positive and* $-x$ *is always negative.*

We would like to define addition, subtraction, multiplication, and division in the integers in such a way that these operations are consistent with those operations over the natural numbers and so that they gain closure, commutativity, associativity, and distributivity under addition and multiplication.

We start with addition. Recall that we defined addition of two natural numbers and of two fractions as the concatenation of two segments. We would like this definition to extend to the integers, but we are confronted with the need to define what a "negative segment" is. This we will do as follows. If $a < 0$, then $-a > 0$. We associate with a a segment of length $|-a|$, but to the left of 0 on the line, as in Figure 2.9.

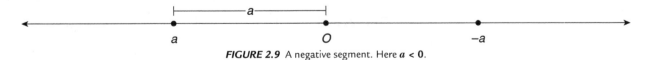

FIGURE 2.9 A negative segment. Here $a < 0$.

So we define the addition of integers a, b, as the concatenation of segments. Depending on the sign of a and b, we have four cases. (We exclude the cases when at least one of a or b is zero, these cases being trivial.)

Example 2.31 (Case $a > 0$, $b > 0$) To add b to a, we first locate a on the line. From there, we move b units right (since $b > 0$), landing at $a + b$. Notice that this case reduces to addition of natural numbers, and therefore, we should obtain the same result as for addition of natural numbers. This example is illustrated in Figure 2.10. For a numerical example (with $a = 3$, $b = 2$), see Figure 2.11.

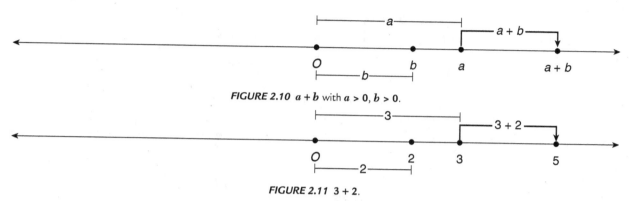

FIGURE 2.10 $a + b$ with $a > 0, b > 0$.

FIGURE 2.11 $3 + 2$.

Example 2.32 (Case $a < 0$, $b < 0$) To add b to a, we first locate a on the line. From there, we move b units left (since $b < 0$), landing at $a + b$. This example is illustrated in Figure 2.12. For a numerical example (with $a = -3$, $b = -2$), see Figure 2.13.

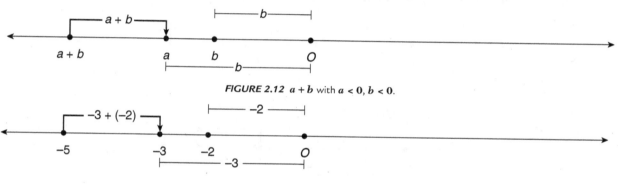

FIGURE 2.12 $a + b$ with $a < 0, b < 0$.

FIGURE 2.13 $-3 + (-2)$.

☞ *Examples 2.31 and 2.32 conform to the following intuitive idea. If we associate positive numbers to "gains" and negative numbers to "losses," then a "gain" plus a "gain" is a "larger gain" and a "loss" plus a "loss" is a "larger loss."*

Example 2.33 We have,

$$(+1) + (+3) + (+5) = +9,$$

since we are adding three gains, and we thus obtain a larger gain.

Example 2.34 We have,

$$(-11) + (-13) + (-15) = -39,$$

since we are adding three losses, and we thus obtain a larger loss.

We now tackle the cases when the summands have opposite signs. In this case, borrowing from the preceding remark, we have a "gain" plus a "loss." In such a case it is impossible to know beforehand whether the result is a gain or a loss. The only conclusion we could gather, again, intuitively, is that the result will be in a sense "smaller," that is, we will have a "smaller gain" or a smaller loss." If the "gain" is larger than the loss, then the result will be a "smaller gain." And if the "loss" is larger than the "gain," then the result will be a "smaller loss."

Example 2.35 (Case $a < 0$, $b > 0$) To add b to a, we first locate a on the line. Since $a < 0$, it is located to the left of O. From there, we move b units right (since $b > 0$), landing at $a + b$. This example is illustrated in Figure 2.14. For a numerical example (with $a = -3$, $b = 2$), see Figure 2.15.

Again, we emphasise, in the sum **(−3) + (+2)**, the "loss" is larger than the "gain." So when adding, we expect a "smaller loss," fixing the sign of the result to be minus.

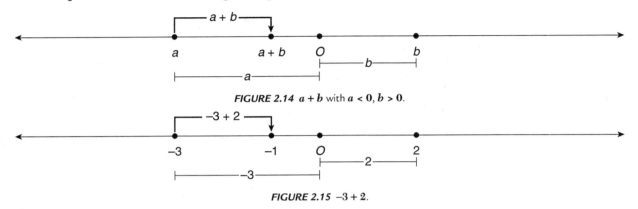

FIGURE 2.14 $a + b$ with $a < 0, b > 0$.

FIGURE 2.15 $-3 + 2$.

Example 3.36 (Case $a > 0$, $b < 0$) To add b to a, we first locate a on the line. From there, we move b units left (since $b < 0$), landing at $a + b$. This example is illustrated in Figure 2.16. For a numerical example (with $a = 3$, $b = -2$), see Figure 2.17.

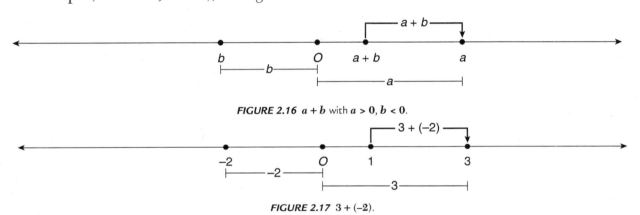

FIGURE 2.16 $a + b$ with $a > 0, b < 0$.

FIGURE 2.17 $3 + (-2)$.

Example 2.37 We have,

$$(+19) + (-21) = -2,$$

since the loss of **21** is larger than the gain of **19**, and so we obtain a loss.

Example 2.38 We have,

$$(-100) + (+210) = +110,$$

since the loss of **100** is smaller than the gain of **210**, and so we obtain a gain.

We now turn to subtraction. We define subtraction in terms of addition.

Definition 2.17 Subtraction is defined as

$$a - b = a + (-b).$$

Example 2.39 We have, $(+8) - (+5) = (+8) + (-5) = 3.$

Example 2.40 We have, $(-8) - (-5) = (-8) + (+5) = -3.$

Example 2.41 We have, $(+8) - (-5) = (+8) + (+5) = 13.$

Example 2.42 We have, $(-8) - (+5) = (-8) + (-5) = -13.$

We now explore multiplication. Again, we would like the multiplication rules to be consistent with those we have studied for natural numbers. This entails that if we multiply two positive integers, the result will be a positive integer. What happens in other cases? Suppose $a > 0$ and $b < 0$. We would like to prove that $ab < 0$. Observe that

$$a(b-b) = 0 \Rightarrow ab - ab = 0 \Rightarrow ab + a(-b) = 0 \Rightarrow ab = -a(-b).$$

Since $-b > 0$, $a(-b)$ is the product of two positive integers, and therefore positive. Thus $-a(-b)$ is negative, and so $ab = -a(-b) < 0$. We have proved that the product of a positive integer and a negative integer is negative. Using the same trick we can prove that

$$(-x)(-y) = xy.$$

If $x < 0$, $y < 0$, then both $-x > 0$, $-y > 0$, so the product of two negative integers is the same as the product of two positive integers, and so it is positive. We have thus proved the following rules:

$$(+)(+) = (-)(-) = + \text{ and } (+)(-) = (-)(+) = -.$$

Intuitively, you may think of a negative sign as a reversal of direction on the real line. Thus the product or quotient of two integers of different signs is negative. Two negatives give two reversals, which is to say, no reversal at all—thus the product or quotient of two integers with the same sign is positive. The sign rules for division are obtained from and are identical to those of division.

Example 2.43 We have,

$$(-2)(5) = -10, \ (-2)(-5) = +10, \ (+2)(-5) = -10, \ (+2)(+5) = +10.$$

Example 2.44 We have,

$$(-20) \div (5) = -4, \ (-20) \div (-5) = +4, \ (+20) \div (-5) = -4, \ (+20) \div (+5) = +4.$$

The rules of operator precedence discussed in the section of natural numbers apply.

Example 2.45 We have,

$$\frac{(-8)(-12)}{3} + \frac{30}{((-2)(3))} = \frac{96}{3} + \frac{30}{(-6)}$$
$$= 32 + (-5)$$
$$= 27.$$

Example 2.46 We have,

$$(5-12)^2 - (-3)^3 = (-7)^2 - (-27)$$
$$= 49 + 27$$
$$= 76.$$

As a consequence of the rule of signs for multiplication, a product containing an odd number of minus signs will be negative and a product containing an even number of minus signs will be positive.

Example 2.47

$$(-2)^2 = 4, \ (-2)^3 = -8, \ (-2)^{10} = 1024.$$

☞ *Notice the difference between, say, $(-a)^2$ and $-a^2$. $(-a)^2$ is the square of $-a$, and so it is always nonnegative. On the other hand, $-a^2$ is the opposite of a^2, and therefore it is always nonpositive.*

Example 2.48 We have,

$$5 + (-4)^2 = 5 + 16 = 21,$$

$$5 - 4^2 = 5 - 16 = -11,$$

$$5 - (-4)^2 = 5 - 16 = -11.$$

Supplementary Exercises

Exercise 2.5.1 *Perform the following operations mentally.*

1. $(-9) - (-17)$
2. $(-17) - (9)$
3. $(-9) - (17)$
4. $(-1) - (2) - (-3)$
5. $(-100) - (101) + (-102)$
6. $|-2| - |-2|$
7. $|-2| - (-|2|)$
8. $|-100| + (-100) - (-(-100))$

Exercise 2.5.2 *Place each of the following nine integers*

$$\{-4, -3, -2, -1, 0, 1, 2, 3, 4\}$$

once in the diagram below so that every diagonal sum is the same.

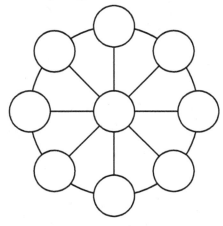

Exercise 2.5.3 *Place each of the nine integers {–2, –1, 0, 1, 2, 3, 4, 5, 6} once in the diagram below so that every diagonal sum is the same.*

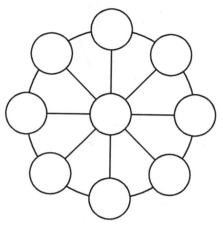

Exercise 2.5.4 *Complete the "crossword" puzzle with* **1**s *or* **–1**s.

–1	×		=	
×	■	×	■	×
	×	–1	=	
=	■	=	■	=
	×		=	–1

Exercise 2.5.5 *Evaluate the expression*

$$\frac{a^3 + b^3 + c^3 - 3abc}{a^2 + b^2 + c^2 - ab - bc - ca}$$

when $a = 2$, $b = -3$, *and* $c = 5$.

2.6 RATIONAL, IRRATIONAL, AND REAL NUMBERS

Definition 2.18 The set of *negative fractions* is the set

$$\left\{ -\frac{a}{b} : a \in \mathbb{N}, a > 0, b \in \mathbb{N}, b > 0 \right\}.$$

The set of positive fractions together with the set of negative fractions and the number **0** form the set of *rational numbers,* which we denote by $\mathbb{Q}$.

The rules for operations with rational numbers derive from those of operations with fractions and with integers. Also, the rational numbers are closed under the four operations of addition, subtraction, multiplication, and division. A few examples follow.

Example 2.49 have

$$\frac{2}{5} \cdot \frac{15}{12} - \frac{7}{10} \div \frac{14}{15} = \frac{2}{5} \cdot \frac{15}{12} - \frac{7}{10} \cdot \frac{15}{14}$$

$$= \frac{\cancel{2}}{\cancel{5}} \cdot \frac{\cancel{5} \cdot \cancel{3}}{\cancel{2} \cdot 2 \cdot \cancel{3}} - \frac{\cancel{7}}{2 \cdot \cancel{5}} \cdot \frac{3 \cdot \cancel{5}}{2 \cdot \cancel{7}}$$

$$= \frac{1}{2} - \frac{3}{4}$$

$$= \frac{2}{4} - \frac{3}{4}$$

$$= -\frac{1}{4}.$$

It can be proved that any rational number has a decimal expansion which is either periodic (repeats) or terminates, and that any number with either a periodic or a terminating expansion is a rational number. For example, $\frac{1}{4} = 0.25$ has a terminating decimal expansion, and $\frac{1}{11} = 0.0909090909... = 0.\overline{09}$ has a repeating one. By long division you may also obtain

$$\frac{1}{7} = 0.\overline{142857}, \qquad \frac{1}{17} = 0.\overline{0588235294117647},$$

and as you can see, the periods may be longer than what your calculator can handle.

What about the numbers whose decimal expansion is infinite and does not repeat? This leads us to the following definition.

Definition 2.19 A number whose decimal expansion is infinite and does not repeat is called an *irrational number*.

From the discussion above, an irrational number is one that cannot be expressed as a fraction of two integers.

Example 2.50 Consider the number

$$0.10100100010000100000100000001...,$$

where the number of **0**s between consecutive **1**s grows in sequence: **1, 2, 3, 4, 5, ...** Since the number of **0**s is progressively growing, this infinite decimal does not have a repeating period and therefore must be an irrational number.

Using my computer, when I enter $\sqrt{2}$ I obtain as an answer

$$1.41421356237309504880168877242097$$

Is this answer exact? Does this decimal repeat? It can be proved that the number $\sqrt{2}$ is irrational, so the above answer is only an approximation and the decimal does not repeat. The first proof of the irrationality of $\sqrt{2}$ is attributed to Hippasus of Metapontum, one of the disciples of Pythagoras (c. 580 b.c.e.– c. 500 b.c.e.). The Greek world view at that time was that all numbers where rational, and therefore this discovery was anathema to the Pythagoreans who decided to drown Hipassus for his discovery.

☞ *It can be proved that if n is a natural number that is not a perfect square, then $\sqrt{n}$ is irrational. Therefore, $\sqrt{2}, \sqrt{3}, \sqrt{5}, \sqrt{6}, \sqrt{7}, \sqrt{8}$, etc., are all irrational.*

In 1760, Johnn Heinrich Lambert (1728 – 1777) proved that n is irrational.

☞ *In particular, then, it would be incorrect to write* $\pi = 3.14$, *or* $\pi = \frac{22}{7}$, *or* $\pi = \frac{355}{113}$, *etc., since* π *is not rational. All of these are simply approximations, and therefore we must write* $\pi \approx 3.14$, *or* $\pi \approx \frac{22}{7}$, *or* $\pi \approx \frac{355}{113}$, *etc.*

Definition 2.20 The set of *real numbers*, denoted by $\mathbb{R}$, is the collection of rational numbers together with the irrational numbers.

Supplementary Exercises

Exercise 2.6.1 *Evaluate the expression*

$$\frac{x}{y+z} + \frac{y}{z+x} + \frac{z}{x+y} - \left(\frac{x}{y} + \frac{y}{z} + \frac{z}{x}\right)^2$$

when $x = -1$, $y = 2$, *and* $z = -3$.

Exercise 2.6.2 *Evaluate the following expressions when* $x = -\frac{2}{3}$ *and* $y = \frac{3}{5}$.

1. $2x + 3y$

2. $xy - x - y$

3. $x^2 + y^2$

Exercise 2.6.3 *In this exercise, you are allowed to use any of the operations* $+$, $-$, $\div$, $\times$, $\cdot$, $!$, *and exponentiation. You must use exactly four 4s. Among your fours you may also use* $.4$. *The* $n!$ *(factorial) symbol means that you multiply all the integers up to* n. *For example,* $1! = 1$, $2! = 1 \cdot 2 = 2$, $3! = 1 \cdot 2 \cdot 3 = 6$, $4! = 1 \cdot 2 \cdot 3 \cdot 4 = 24$. *With these rules, write every integer, from 1 to 20 inclusive. For example,*

$$11 = \frac{4}{.4} + \frac{4}{4}, \quad 15 = \frac{44}{4} + 4, \quad 20 = \frac{4}{.4} + \frac{4}{.4}, \quad 13 = 4! - \frac{44}{4}.$$

Exercise 2.6.4 *Without using any of the signs* $+$, $-$, $\div$, $\cdot$, *but with exponentiation being allowed, what is the largest number that you can form using three 4s? Again, you must explain your reasoning.*

Exercise 2.6.5 *Find the value of* $xy = \frac{3}{5}$.

Exercise 2.6.6 *Suppose you know that* $\frac{1}{3} = 0.333333... = 0.\overline{3}$. *What should* $0.1111... = 0.\overline{1}$ *be?*

Exercise 2.6.7 *Find the value of* $121(0.\overline{09})$.

Exercise 2.6.8 *Use a calculator to round* $121(0.\overline{09})$. *to two decimal places.*

Exercise 2.6.9 *Use a calculator to round* $\sqrt{2} \cdot \sqrt{3} \cdot \sqrt{5}$ *to two decimal places.*

Exercise 2.6.10 *Let* a, b *be positive real numbers. Is it always true that* $\sqrt{a+b} = \sqrt{a} + \sqrt{b}$?

ALGEBRAIC OPERATIONS

ADDITION AND SUBTRACTION

This chapter illustrates the mathematical techniques of addition and subtraction of algebraic equations, with an emphasis on *Terms and Algebraic Expressions* and *More Suppression of Parentheses*.

3.1 TERMS AND ALGEBRAIC EXPRESSIONS

If one classroom has a group of **7** freshmen students, **8** sophomore students, and **3** junior students—and another classroom has **4** freshmen students, **4** sophomore students, and **1** junior student—we could write the following equation addressing the total number of students:

$$(7F + 8S + 3J) + (4F + 4S + 1J) = 11F + 12S + 4J.$$

The procedure of *collecting like terms*, which we will explain shortly, draws essentially from the concept used in this example.

Again, consider the following way of adding **731** and **695**. Since

$$731 = 7 \cdot 10^2 + 3 \cdot 10 + 1, \quad 695 = 6 \cdot 10^2 + 9 \cdot 10 + 5,$$

we could add in the following fashion, without worrying about carrying:

$$(7 \cdot 10^2 + 3 \cdot 10 + 1) + (6 \cdot 10^2 + 9 \cdot 10 + 5) = 13 \cdot 10^2 + 12 \cdot 10 + 6 = 1300 + 120 + 6 = 1426.$$

Definition 3.1 An *algebraic expression* is a collection of symbols (letters and/or numbers). An algebraic expression may have one or more *terms*, which are separated from each other by the signs + (plus) or − (minus).

Example 3.1 The expression $18a + 3b - 5$ consists of three terms.

Example 3.2 The expression $18ab^2c$ consists of one term. Notice in this case that since no sign precedes $18ab^2c$, the + sign is tacitly understood. In other words, $18ab^2c = + 18ab^2c$.

Example 3.3 The expression $a + 3a^2 - 4a^3 - 8ab + 7$ consists of five terms.

Definition 3.2 When one of the factors of a term is a number, it is called the *numerical coefficient* (or *coefficient*, for short) of the term. In the expression a^b, a is the base and b is the exponent.

Example 3.4 In the expression $13a^2b^3$, **13** is the numerical coefficient of the term, **2** is the exponent of a and **3** is the exponent of b.

Example 3.5 In the expression a^2b^d, 1 is the numerical coefficient of the term, **2** is the exponent of a and d is the exponent of b.

Example 3.6 In the expression $-a^c$, -1 is the numerical coefficient of the term and c is the exponent of a. Notice then that the expression $-1a^c$ is equivalent to the expression $-a^c$.

Definition 3.3 If two terms

1. agree in their letters, and

2. for each letter appearing in them the exponent is the same,

then we say that the terms are *like terms*.

Example 3.7 The terms $-a, 7a, 5a$, are all like terms.

Example 3.8 The terms $4a^2b^3$, $7b^3a^2$, $5a^2b^3$, are all like terms. Notice that by commutativity of multiplication, $7b^3a^2 = 7a^2b^3$.

Example 3.9 The terms $4a$ and $4a^2$ are unlike terms, since the exponent of a in $4a$ is 1 and the exponent of a in $4a^2$ is 2.

Example 3.10 The terms $4ab^2$ and $4ab^3$ are unlike terms, since the exponent of b in $4ab^2$ is 2 and the exponent of b in $4ab^3$ is 3.

To add or collect like terms, we simply add the coefficients of the like terms.

Example 3.11 We have,

$$(a + b) + (a + b) + (a + b) = a + a + a + b + b + b = 3a + 3b$$

☞ *By the commutative law, we may also write this as $3b + 3a$.*

Example 3.12 We have,

$$(-8a) + 9a + (-15a) + 4a = a + (-15a) + 4a$$
$$= (-14a) + 4a$$
$$= -10a$$

Example 3.13 We have,

$$(5\clubsuit - 7\spadesuit) + (-5\clubsuit - 7\spadesuit) = (5\clubsuit) + (-5\clubsuit) + (-7\spadesuit) + (-7\spadesuit)$$
$$= -14\spadesuit.$$

Example 3.14 We have,

$$\frac{5a}{7} + \frac{(-6a)}{7} + \frac{(-18a)}{7} = \frac{5a + (-6a) + (-18a)}{7} = \frac{-19a}{7}.$$

Example 3.15 We have,

$$\frac{5a}{7} + \frac{6a}{7} + \frac{(-18a)}{7} = \frac{5a + 6a + (-18a)}{7} = \frac{-7a}{7} = -a.$$

Example 3.16 We have,

$$(-13a + 5b) + (18a + (-18b)) = -13a + 18a + 5b + (-18b)$$
$$= 5a + (-13b)$$
$$= 5a - 13b$$

☞ *By the commutative law, we may also write this as* **$-13b + 6a$.**

Example 3.17 We have,

$$(5 - 4x + 3x^2) + (-1 - 8x - 3x^2) = 5 + (-1) - 4x + (-8x) + 3x^2 + (-3x^2)$$

$$= 4 + (-12x) + 0$$

$$= 4 + (-12x)$$

$$= 4 - 12x.$$

☞ *By the commutative law, we may also write this as* **$-12x + 4$.**

Example 3.18 We have,

$$(x^2 + xy + y^2 + xy^2) + (-x^2 + 2yx - 3y^2 - x^2y) = 3xy - 2y^2 + xy^2 - x^2y.$$

This last expression is one of the many possible ways to write the final result, thanks to the commutative law.

Supplementary Exercises

Exercise 3.1.1 *A boy buys **a** marbles, wins **b** marbles, and loses **c** marbles. How many marbles does he have? Write one or two sentences explaining your solution.*

Exercise 3.1.2 *Collect like terms* mentally.

1. $(-9a) + (+17a) - 8a$

2. $(-17a) + (-9a) + 20a$

3. $-a - 2a - 3a - 4a$

4. $10a - 8a + 6a - 4a + 2a$

5. $-5a - 3b - 9a - 10b + 10a + 17b$

Exercise 3.1.3 *Collect like terms:*

$$(a + b^2 + c + d^2) + (a^2 - b - c^2 - d) + (a^2 - b^2 + c - d) + (a - b - c^2 - d^2).$$

Exercise 3.1.4 *Collect like terms.*

1. $(a + 2b + 3c) + (3a + 2b + c)$

2. $(a + 2b + 3c) + (3a - 2b - c)$

3. $(a + 2b + 3c) + (3a + 2b + c) + (a + b + c)$

4. $(a + b + c) + (a - b + c) + (-a - b - c)$

5. $(x^2 - 2x + 1) + (x^2 + 2x + 1)$

6. $(x^3 - x^2 + x - 1) + (2x^2 - x + 2)$

7. $(2x^3 - 2x^2 - 2) + (2x^3 - 2x - 1) + (2x^3 - x^2 - x + 1)$

8. $\dfrac{(-5a)}{8} + \dfrac{19a}{8} + \dfrac{(-23a)}{8}$

Exercise 3.1.5 *Find the value of the expression*

$$(2x^2 + x - 2) + (-x^2 - x + 2)$$

when $x = 2$.

Exercise 3.1.6 *Is it always true that* $a + a = 2a$?

Exercise 3.1.7 *Is it always true that* $a + a = 2a^2$?

Exercise 3.1.8 *Is it always true that* $a + a^2 = 2a^3$?

Exercise 3.1.9 *If* $(8x - 5)5 = Ax^5 + Bx^4 + Cx^3 + Dx^2 + Ex + F$, *find* $A + B + C + D + E + F$.

Exercise 3.1.10 *Give three different equivalent forms for the expression* $3yx - x^2\, y$.

3.2 MORE SUPPRESSION OF PARENTHESES

A minus sign preceding an expression in parentheses negates every term of the expression. Thus

$$-(a + b) = -a - b.$$

Example 3.19 We have,

$$(a - 2b) - (-a + 5b) = a - 2b + a - 5b$$
$$= a + a - 2b - 5b$$
$$= 2a - 7b.$$

Example 3.20 We have,

$$(3 - 2x - x^2) - (5 - 6x - 7x^2) = 3 - 2x - x^2 - 5 + 6x + 7x^2$$
$$= 3 - 5 - 2x + 6x - x^2 + 7x^2$$
$$= -2 + 4x + 6x^2.$$

Example 3.21 We have,

$$(a - b) - (2a - 3b) - (-a + 3b) = a - b - 2a + 3b + a - 3b$$
$$= a - 2a + a - b + 3b - 3b$$
$$= -b.$$

Example 3.22 We have,

$$(x^2 - x - 1) - (-x - 3x^2) + (2x^2 - 8) = x^2 - x - 1 + x + 3x^2 + 2x^2 - 8$$
$$= x^2 + 3x^2 + 2x^2 - x + x - 1 - = 8$$
$$= 6x^2 - 9.$$

Recall that the distributive law states that for real numbers $a, b, c,$

$$a(b + c) = ab + ac.$$

We now apply this to algebraic expressions.

Example 3.23 Multiply term by term: $2(a + 2b - 3c)$.

Solution: ▶ *We have,*

$$2(a + 2b - 3c) = 2a + 2(2b) + 2(-3c) = 2a + 4b - 6c.$$

◀

Example 3.24 Multiply term by term: $-2(a + 2b - 3c)$.

Solution: ▶ *We have,*

$$-2(a + 2b - 3c) = -2a + (-2)(2b) + (-2)(-3c) = -2a - 4b + 6c.$$ ◀

Example 3.25 Divide term by term: $\frac{2a - 4b + 7c}{2}$.

Solution: ▶ *We have,*

$$\frac{2a - 4b + 7c}{2} = \frac{2a}{2} + \frac{-4b}{2} + \frac{7c}{2} = a - 2b + \frac{7c}{2}.$$ ◀

☞ *We prefer to use improper fractions and write $\frac{7c}{2}$ rather than $3\frac{1}{2}c$. Also, the expressions $\frac{7}{2}c$ and $\frac{7c}{2}$ are identical, that is, $\frac{7}{2}c = \frac{7c}{2}$.*

We may combine the distributive law and combining like terms.

Example 3.26 Combine like terms: $2(a - 2b) + 3(-a + 2b)$.

Solution: ▶ *We have,*

$$2(a - 2b) + 3(-a + 2b) = 2a - 4b + (-3a) + 6b$$
$$= -a + 2b.$$ ◀

Example 3.27 Combine like terms: $2(a - 2b) - 3(-a + 2b)$.

Solution: ▶ *We have,*

$$2(a - 2b) - 3(-a + 2b) = 2a - 4b + 3a - 6b$$
$$= 5a - 10b.$$ ◀

Example 3.28 Combine like terms: $-3(1 - 2x + 4x^2) + \frac{9x^2 - 3x + 6}{-3}$.

Solution: ▶ *We have,*

$$-3(1 - 2x + 4x^2) + \frac{9x^2 - 3x + 6}{-3} = -3 + 6x - 12x^2 - 3x^2 + x - 2$$
$$= -15x^2 + 7x - 5.$$ ◀

Example 3.29 Combine like terms: $\frac{3}{2}\left(\frac{x}{3} - 2\right) - 2\left(\frac{x}{12} + 2\right)$.

Solution: ▶ *We have,*

$$\frac{3}{2}\left(\frac{x}{3} - 2\right) - 2\left(\frac{x}{12} + 2\right) = \frac{3x}{6} - \frac{6}{2} - \frac{2x}{12} - 4$$
$$= \frac{6x}{12} - \frac{2x}{12} - 3 - 4$$
$$= \frac{4x}{12} - 7$$
$$= \frac{x}{3} - 7.$$ ◀

Supplementary Exercises

Exercise 3.2.1 *A rod measuring $a + b$ inches long was cut by $b - c$ inches. How much of the rod remains? Write one or two sentences explaining your answer.*

Exercise 3.2.2 *A group of x people is going on an excursion and decides to charter a plane. The total price of renting the plane is D dollars, and everyone will pay an equal share. How much does each person pay for their share? On the day of departure, p people are no-shows. How much more will the people who show up have to pay in order to cover the original cost?*

Exercise 3.2.3 *Suppress parentheses and collect like terms:*

$$(a + b^2 + c + d^2) - (a^2 - b - c^2 - d) + (a^2 - b^2 + c - d) - (a - b - c^2 - d^2).$$

Exercise 3.2.4 *Expand and simplify:*

$$2(a + b - 2c) - 3(a - b - c).$$

Exercise 3.2.5 *Suppress parentheses and collect like terms.*

1. $(a + 2b + 3c) - (3a + 2b + c)$

2. $(a + 2b + 3c) + (3a + 2b + c) - (a + b + c)$

3. $(a + b + c) - (a - b + c) - (-a - b - c)$

4. $(x^3 - x^2 + x - 1) - (2x^2 - x + 2)$

5. $(2x^3 - 2x^2 - 2) - (2x^3 - 2x - 1) - (2x^3 - x^2 - x + 1)$

6. $(8.1a - 9.2b) - (-0.2a - 3.1b) - (-a + 3b)$

7. $-\left(-\dfrac{14b}{a} + \dfrac{15b}{a} - \dfrac{1}{a}\right) - \left(-\dfrac{4b}{a} + \dfrac{5}{a} - \dfrac{8b}{a}\right)$

8. $-\left(\dfrac{4}{a} - \dfrac{5}{a} - 2\right) - \left(\dfrac{4}{a} + \dfrac{5}{a} - 8\right)$

9. $(5\clubsuit - 7\spadesuit) - (-5\clubsuit - 7\spadesuit)$

Exercise 3.2.6 *Collect like terms and write in a single fraction:*

$$\frac{a}{3} - \frac{a^2}{5} + \frac{a}{2} + \frac{5a^2}{6}.$$

Exercise 3.2.7 *Combine like terms.*

1. $2(-x + 2y) + 3(x - 2y) - (x + y) + (y - 2x)$

2. $\dfrac{14x - 7}{7} - \dfrac{1}{3}(-3x + 12)$

3. $\dfrac{4x - 12x^2 + 6}{-2} - 2(1 - 2x + x^2)$

Exercise 3.2.8 *Match each expression on the left with the equivalent expression on the right.*

_____ 1.	$x - y$	A.	$y - x$
_____ 2.	$x + y$	B.	$-y + x$
_____ 3.	$-(x - y)$	C.	$x - (-y)$

Exercise 3.2.9 *Mary starts the day with* A *dollars. Her uncle Bob gives her enough money to double her amount. Her aunt Rita gives Mary* **10** *dollars. Mary has to pay* **B** *dollars in fines, and spent* **12** *dollars fueling her car with gas. How much money does she has left?*

Exercise 3.2.10 *Simplify each of the following expressions as much as possible.*

1. $(2-4)x - 5(-6)x$

2. $5 - 2(2x)$

3. $(5-2)(1-2^2)^3 x$

Exercise 3.2.11 *Remove all redundant parentheses. If removing parentheses changes the expression, explain.*

1. $(6t) + x$

2. $6(t+x)$

3. $(ab)^2$

4. $(ab-c)^2$

Exercise 3.2.12 *Fill in the box to make the expression true.*

1. $\dfrac{a-2b}{c} = \dfrac{a}{c} + \boxed{}$

2. $\dfrac{t+3}{t-3} = (t+3)\boxed{}$

Exercise 3.2.13 *Fill in the box to make the expression true.*

1. $1 - t = \boxed{}\, t + 1$

2. $3x + y - t = \boxed{} + 3x$

Exercise 3.2.14 *What is the opposite of* $4 - y + 2$?

Exercise 3.2.15 *What is the additive inverse of* $\dfrac{-a}{-(b-c)}$?

CHAPTER 4

MULTIPLICATION

This chapter illustrates the techniques used to perform multiplication operations in algebra. We introduce mathematical operations in the following sections: *Laws of Exponents, Negative Exponents, Distributive Law, Square of a Sum, Difference of Squares, Cube of a Sum,* and *Sum and Difference of Cubes.*

4.1 LAWS OF EXPONENTS

Recall that if a is a real number and if n is a natural number, then

$$a^n = \underbrace{a \cdot a \cdots a,}_{n\ a's}$$

with the interpretation that $a^0 = 1$ if $a \neq 0$.

Suppose we wanted to compute $a^2 a^5$. We would proceed as follows:

We know that $a^2 = aa$ and $a^5 = aaaaa$, thus:

$$(a^2)(a^5) = (aa)(aaaaa) = aaaaaaa = a^7,$$

The above example can be generalized as the following theorem.

Theorem 4.1 (First Law of Exponents) Let a be a real number and m, n natural numbers. Then

$$a^m a^n = a^{m+n}$$

Example 4.1 $(ab^2c^3)(a^3b^3c) = a^{1+3}b^{2+3}c^{3+1} = a^4b^5c^4$.

Example 4.2 $(a^x)(b^{2y})(a^{2x}b^{2y}) = a^{x+2x}b^{2y+2y} = a^{3x}b^{4y}$.

Example 4.3 $(a^{x-2y+z})(a^{2x-y-z})(a^{x+y+z}) = a^{x-2y+z+2x-y-z+x+y+z} = a^{4x-2y+z}$.

Example 4.4 $5^5 + 5^5 + 5^5 + 5^5 + 5^5 = 5(5^5) = 5^{1+5} = 5^6 = 15625$.

We may use associativity and commutativity in the case when the numerical coefficients are different from 1.

Example 4.5 $(5x^2)(6x^3) = 30x^5$.

Example 4.6 $(-2ab^2c^3)(3abc^4) = -6a^2b^3c^7$.

We now tackle division. As a particular case, observe that

$$\frac{a^5}{a^2} = \frac{aaaaa}{aa} = \frac{aaa\cancel{aa}}{\cancel{aa}} = a^3,$$

and, $a^3 = a^{5-2}$. This generalizes as the following theorem.

Theorem 4.2 (Second Law of Exponents) Let $a \neq 0$ be a real number and m, n natural numbers, such that $m \geq n$. Then

$$\frac{a^m}{a^n} = a^{m-n}.$$

Example 4.7 $\frac{2^9}{2^4} = 2^{9-4} = 2^5 = 32.$

Example 4.8 $\frac{-24a^{12}b^9c^5d^2}{2a^6b^3c^4d^2} = -12a^{12-6}b^{9-3}c^{5-4}d^{2-2} = -12a^6b^6c^1d^0 = -12a^6b^6c.$

Example 4.9 $\frac{(5^a)(5^b)}{5^{a+b-2}} = 5^{a+b-(a+b-2)} = 5^2 = 25.$

Example 4.10 $\frac{15x^5y^7z^4}{5x^2y^2z^2} = 3x^3y^5z^2.$

Example 4.11 $\frac{16b^2yx^2}{-2xy} = -8b^2x.$

Example 4.12 $\frac{7a^2bc}{-7a^2bc} = -1.$

Example 4.13 $\frac{(a+b)^4(2+a+b)^6}{(a+b)^3(2+a+b)^4} = (a+b)(2+a+b)^2$

Example 4.14 $\frac{a^5b^4(a+b)^3}{a^3b^4(a+b)} = a^2(a+b)^2.$

Suppose we wanted to compute $(a^2)^5$. Since a quantity raised to the fifth power is simply the quantity multiplied by itself five times, we have:

$$(a^2)^5 = (a^2)(a^2)(a^2)(a^2)(a^2) = a^{2+2+2+2+2} = a^{10},$$

upon using the first law of exponents. In general we have the following law dealing with "exponents of exponents."

Theorem 4.3 (Third Law of Exponents) Let a be a real number and m, n natural numbers. Then

$$(a^m)^n = a^{mn}$$

Theorem 4.4 (Fourth Law of Exponents) Let a and b be real numbers and let m be a natural number. Then

$$(ab)^m = a^m b^m.$$

Example 4.15 We have,

$$(2ab^2c^3)^2(-3a^3bc^2)^3 = (2^2(a^1)^2(b^2)^2(c^3)^2)((-3)^3(b^1)^3(c^2)^3)$$

$$= (4a^2b^4c^6)(-27a^9b^3c^6)$$

$$= -108a^{11}b^7c^{12}.$$

Example 4.16 $(a^{x^3})^{2x} = a^{(x^3)(2x)} = a^{2x^4}.$

Example 4.17 $25^3 4^3 = (25 \cdot 4)^3 = 100^3 = 1000000.$

Supplementary Exercises

Exercise 4.1.1 *Simplify:* $\frac{24xyz^3}{-3z^2}$.

Exercise 4.1.2 *Simplify:* $\frac{-168a^2b^2cx^2}{-7abx^2}$.

Exercise 4.1.3 *Find the exact numerical value of:* $\frac{20^6}{20^5}$.

Exercise 4.1.4 *Find the exact numerical value of:* $\frac{25^2 + 25^2 + 25^2 + 25^2}{25^4}$.

Exercise 4.1.5 *If*

$$\left(\frac{4^5 + 4^5 + 4^5 + 4^5}{3^5 + 3^5 + 3^5}\right)\left(\frac{6^5 + 6^5 + 6^5 + 6^5 + 6^5 + 6^5}{2^5 + 2^5}\right) = 2^n,$$

*find **n**.*

Exercise 4.1.6 *If*

$$3^{2001} + 3^{2002} + 3^{2003} = a3^{2001},$$

*find **a**.*

Exercise 4.1.7 *Demonstrate that:*

$$\left(a^x b^y\right)\left(\frac{b^{2x}}{a^{-y}}\right) = a^{x+y}b^{y+2x}.$$

Exercise 4.1.8 *Divide:* $\frac{1}{x} \div \frac{1}{x^2}$.

Exercise 4.1.9 *Multiply:*

$$\left(\frac{1}{x}\right)\left(\frac{1}{x^2}\right).$$

Exercise 4.1.10 *Divide:*

$$\frac{1}{s-1} \div \frac{1}{(s-1)^2}.$$

Exercise 4.1.11 *Multiply:*

$$\left(\frac{1}{s-1}\right)\left(\frac{1}{(s-1)^2}\right).$$

Exercise 4.1.12 *Divide:*

$$\frac{a^2}{x} \div \frac{x}{a^2}.$$

Exercise 4.1.13 *Multiply:*

$$\left(\frac{a^2}{x}\right)\left(\frac{x}{a^2}\right).$$

Exercise 4.1.14 *Simplify and write as a single number:*

$$\frac{6^5 + 6^5 + 6^5 + 6^5 + 6^5 + 6^5 + 6^5 + 6^5 + 6^5 + 6^5}{3^3 + 3^3 + 3^3 + 3^3 + 3^3}.$$

Exercise 4.1.15 *If* $\frac{a^8}{b^9} \div \frac{(a^2 b)^3}{b^{20}} = a^m b^n,$ *then* $m =$ _____ *and* $n =$ _____.

4.2 NEGATIVE EXPONENTS

Let $a \neq 0$ be a real number and let n be a natural number. Then

$$1 = a^0 = a^{n-n} = a^n a^{-n} \Rightarrow a^{-n} = \frac{1}{a^n}.$$

This gives an interpretation to negative exponents. Observe also that taking reciprocals,

$$a^{-n} = \frac{1}{a^n} \Rightarrow \frac{1}{a^{-n}} = \frac{1}{\frac{1}{a^n}} = a^n.$$

Example 4.18 $2^{-3} = \frac{1}{2^3} = \frac{1}{8}.$

Example 4.19 $\frac{2^4}{2^9} = 2^{4-9} = 2^{-5} = \frac{1}{2^5} = \frac{1}{32}.$

Example 4.20 $\frac{2^{-3}}{3^{-2}} = \frac{3^2}{2^3} = \frac{9}{8}.$

Example 4.21 $\left(\frac{2^4}{5^{-3}}\right)^{-2} = \frac{2^{-8}}{5^6} = \frac{1}{256.15625} = \frac{1}{4000000}.$

Example 4.22 Simplify, using positive exponents only:

$$\frac{x^{-6}}{y^9} \cdot \frac{y^{-8}}{x^{-12}}.$$

Solution: ▶

$$\frac{x^{-6}}{y^9} \cdot \frac{y^{-8}}{x^{-12}} = \frac{x^{-6} y^{-8}}{y^9 x^{-12}} = x^{-6-(-12)} y^{-8-9} = x^6 y^{-17} = \frac{x^6}{y^{17}}.$$

◀

Example 4.23 Simplify, using positive exponents only:

$$\left(\frac{x^{-6}}{y^9}\right)^{-3} \cdot \left(\frac{y^{-1}}{x^{-2}}\right)^2.$$

Solution: ▶ *We have,*

$$\left(\frac{x^{-6}}{y^9}\right)^{-3} \cdot \left(\frac{y^{-1}}{x^{-2}}\right)^2 = \frac{x^{18}}{y^{-27}} \cdot \frac{y^{-2}}{x^{-4}} = x^{18-(-4)} y^{-2-(-27)} = x^{22} y^{25}.$$

◀

Supplementary Exercises

Exercise 4.2.1 *Simplify, using positive exponents only.*

1. $\dfrac{x^{-5}}{x^4}$

2. $\dfrac{x^4}{x^{-5}}$

3. $x^{-5}x^4$

4. $(x^{-1}y^2z^3)^3$

5. $\dfrac{a^3a^{-2}}{a^{-6}b^4}\,3$

6. $\dfrac{(ab)(ab^2)}{(a^{-1}b^{-2})^2(a^2b^2)}$

Exercise 4.2.2 *Demonstrate that* $(a^{x+3}b^{y-4})(a^{2x-3}b^{-2y+5}) = a^{3x}a^{1-y}$.

Exercise 4.2.3 *Demonstrate that*

$$\frac{a^x b^y}{a^{2x}b^{-2y}} = a^{-x}b^{3y}.$$

Exercise 4.2.4 *Simplify and express with positive exponents only:* $\dfrac{x^6}{y^9} \div \dfrac{x^{-2}}{y^{-3}}$

Exercise 4.2.5 *Calculate the exact numerical value:*

$$\frac{2^{20}6^{10}}{4^{11}3^{10}}.$$

Exercise 4.2.6 *Calculate the exact numerical value:*

$$\left(\left(\frac{2}{5}\right)^{-1} - \left(\frac{1}{2}\right)^{-1} \right)^{-2}.$$

Exercise 4.2.7 *Calculate the exact numerical value:*

$$\frac{(2^4)^8}{(4^8)^2}.$$

4.3 DISTRIBUTIVE LAW

Recall that the distributive law says that if **a, b,** and **c** are real numbers, then

$$a(b+c) = ab + ac \qquad (a+b)c = ac + bc.$$

We now apply the distributive law to multiplication of terms.

Example 4.24 We have,

$$-3xy^2(2x^3 - 5xy^3) = (-3xy^2)(2x^3) - (-3xy^2)(5xy^3)$$

$$= -6x^4y^2 + 15x^2y^5.$$

Example 4.25 We have,

$$(-3a^2b)(-3ab^2 + 2b) - (a^2 - 3a^2b)(ab^2) = 9a^3b^3 - 6a^2b^2 - (a^3b^2 - 3a^3b^3).$$

$$= 9a^3b^3 - 6a^2b^2 + (-a^3b^2) + 3a^3b^3$$

$$= 12a^3b^3 - 6a^2b^2 - a^3b^2.$$

From the distributive law we deduce that

$$(a + b)(c + d) = (a + b)c + (a + b)d = ac + bc + ad + bd.$$

Example 4.26 We have,

$$(2x + 3)(5x + 7) = (2x + 3)(5x) + (2x + 3)(7)$$

$$= 10x^2 + 15x + 14x + 21$$

$$= 10x^2 + 29x + 21.$$

☞ *Because of commutativity, we may also proceed as follows:*

$$(2x + 3)(5x + 7) = (2x)(5x + 7) + 3(5x + 7)$$

$$= 10x^2 + 14x + 15x + 21$$

$$= 10x^2 + 29x + 21.$$

Example 4.27 We have,

$$(2x - 3)(x - 4) = (2x - 3)(x) + (2x - 3)(-4)$$

$$= 2x^2 - 3x - 8x + 12$$

$$= 2x^2 - 11x + 12.$$

☞ *Because of commutativity, we may also proceed as follows:*

$$(2x - 3)(x - 4) = (2x)(x - 4) - 3(x - 4)$$

$$= 2x^2 - 8x - 3x + 12$$

$$= 2x^2 - 11x + 12.$$

We also recall the meaning of exponentiation.

Example 4.28 We have,

$$(x - 4)^2 = (x - 4)(x - 4)$$

$$= x(x - 4) - 4(x - 4)$$

$$= x^2 - 4x - 4x + 16$$

$$= x^2 - 8x + 16.$$

Example 4.29 To expand $(x + 1)^3$, we first find

$$(x + 1)^2 = (x + 1)(x + 1) = x(x + 1) + 1(x + 1) = x^2 + x + x + 1 = x^2 + 2x + 1.$$

Then we use this result to find the following:

$$(x + 1)^3 = (x + 1)(x + 1)^2$$
$$= (x + 1)(x^2 + 2x + 1)$$
$$= x(x^2 + 2x + 1) + 1(x^2 + 2x + 1)$$
$$= x^3 + 2x^2 + x + x^2 + 2x + 1$$
$$= x^3 + 3x^2 + 3x + 1.$$

Supplementary Exercises

Exercise 4.3.1 *Expand and simplify.*

1. $x(x^2 + x + 1)$

2. $2x(x^2 - 2x + 3)$

3. $(2x + 1)(2x + 1)$

4. $(2x + 1)(2x - 1)$

5. $(2x - 1)(2x - 1)$

Exercise 4.3.2 *Expand and simplify:*

$$(x^2 + 1) + 2x(x^2 + 2) - 3x^3(x^2 + 3).$$

Exercise 4.3.3 *Expand and simplify:*

$$4x^3(x^2 - 2x + 2) - 2x^2(x^2 - x^2 - 1).$$

Exercise 4.3.4 *Multiply and collect like terms.*

1. $-2x(x + 1) + 3x^2(x - 3x^2)$

2. $-2x(x + 1)^2$

3. $a^2c(ab - ac) + ac^2(a^2 - 2b)$

4. $(x + 2y + 3z)(x - 2y)$

Exercise 4.3.5 *Expand the product:* $(2x + 2y + 1)(x - y + 1).$

Exercise 4.3.6 *Expand and collect like terms:* $(x + 2)(x + 3) - (x - 2)(x - 3).$

Exercise 4.3.7 *Expand and collect like terms:* $(x - 1)(x^2 + x + 1) - (x + 1)(x^2 - x + 1).$

Exercise 4.3.8 *Expand and simplify:* $(x + y - z)(x - y + 2z).$

Exercise 4.3.9 *Expand the product:* $(a + b + c)(a^2 + b^2 + c^2 - ab - bc - ca).$

Exercise 4.3.10 *Expand the product:* $(a + b + c)(x + y + z).$

Exercise 4.3.11 *Prove that the product of two even integers is even.*

Exercise 4.3.12 *Prove that the product of two odd integers is odd.*

Exercise 4.3.13 *Prove that the product of two integers that leave remainder* **3** *upon division by* **4** *leaves remainder* **1** *upon division by* **4**.

Exercise 4.3.14 *Prove that the product of any two integers leaving remainder* **2** *upon division by* **3** *leaves remainder* **1** *upon division by* **3**.

Exercise 4.3.15 *Prove that the product of two integers leaving remainder **2** upon division by **5** leaves remainder **4** upon division by **5**.*

Exercise 4.3.16 *Prove that the product of two integers leaving remainder **3** upon division by **5** leaves remainder **4** upon division by **5**.*

Exercise 4.3.17 *If $x^2 + x - 1 = 0$, find $x^4 + 2x^3 + x^2$.*

Exercise 4.3.18 *Expand and collect like terms:*

$$2x(x + 1) - x^2 - x(x - 1).$$

Then use your result to evaluate

$$2 \cdot (11112) \cdot (11113) - 11112^2 - (11111) \cdot (11112).$$

without a calculator.

4.4 SQUARE OF A SUM

In this section, we study some special products that often occur in algebra.

Theorem 4.5 (Square of a Sum) For all real numbers a, b the following identity holds

$$(a \pm b)^2 = a^2 \pm 2ab + b^2.$$

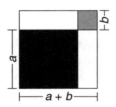

FIGURE 4.1 $(a + b)^2 = a^2 + 2ab + b^2$

The identity $(a + b)^2 = a^2 + 2ab + b^2$ has a graphical justification in terms of areas of squares, as seen in Figure 4.1. The area of the large $(a + b) \times (a + b)$ square is $(a + b)^2$. In terms, this area can be divided into two square regions, one of area a^2 and the other of area b^2, and two rectangular regions each of area ab.

Example 4.30 We have,

$$(2x + 3y)^2 = (2x)^2 + 2(2x)(3y) + (3y)^2 = 4x^2 + 12xy + 9y^2.$$

Example 4.31 We have,

$$(2 - x^2)^2 = (2)^2 - 2(2)(x^2) + (x^2)^2 = 4 - 4x^2 + x^4.$$

The principle illustrated here can be used to facilitate mental arithmetic.

Example 4.32 One may compute mentally 52^2 as follows:

$$52^2 = (50 + 2)^2 = 50^2 + 2(50)(2) + 2^2 = 2500 + 200 + 4 = 2704.$$

The formula $(a + b)^2 = a^2 + 2ab + b^2 = a^2 + b^2 + 2ab$ gives the square of a sum in terms of the sum of squares and the product of the numbers. Thus knowing any two of them will determine the other.

Example 4.33 If the sum of two numbers is **3** and their product **4**, find the sum of their squares.

Solution: ▶ *Let the two numbers be* x, y. *Then* $x + y = 3$ *and* $xy = 4$. *Therefore*

$$9 = (x + y)^2 = x^2 + 2xy + y^2 = x^2 + 8 + y^2 \Rightarrow x^2 + y^2 = 1. \qquad ◀$$

Supplementary Exercises

Exercise 4.4.1 *Expand and simplify:* $(3 - y)^2$.

Exercise 4.4.2 *Expand and simplify:* $(5 - 2x^2)^2$.

Exercise 4.4.3 *Expand and simplify:* $(2ab^2 - 3c^3d^4)^2$.

Exercise 4.4.4 *Expand and simplify:* $(x + 2y)^2 + 4xy$.

Exercise 4.4.5 *Expand and simplify:* $(x + 2y)^2 - 4xy$.

Exercise 4.4.6 *Expand:* $(xy - 4y)^2$.

Exercise 4.4.7 *Expand:* $(ax + by)^2$.

Exercise 4.4.8 *Expand and simplify:* $(a + 2)^2 + (a - 2)^2$.

Exercise 4.4.9 *Expand and simplify:* $(a + 2)^2 - (a - 2)^2$.

Exercise 4.4.10 *Expand and simplify:* $(a + 2b + 3c)^2$.

Exercise 4.4.11 *Expand and simplify:* $(a + 2b - c)^2$.

Exercise 4.4.12 *If* $x + \frac{1}{x} = 6$, *find* $x^2 + \frac{1}{x^2}$.

Exercise 4.4.13 *Find the value of* n *in* $(10^{2002} + 25)^2 - (10^{2002} - 25)^2 = 10^n$.

Exercise 4.4.14 *Given that* $2x + 3y = 3$ *and* $xy = 4$, *find* $4x^2 + 9y^2$.

Exercise 4.4.15 *If the difference of two numbers is* 3 *and their product* 4, *find the sum of their squares.*

Exercise 4.4.16 *Given that* $x + y = 4$ *and* $xy = -3$, *find* $x^2 + y^2$.

Exercise 4.4.17 *Given that* $x + y = 4$ *and* $xy = -3$, *find* $x^4 + y^4$.

Exercise 4.4.18 *Expand:* $(x + y + z)^2$.

Exercise 4.4.19 *Demonstrate that*

$$(x + y + z + w)^2 = x^2 + y^2 + z^2 + w^2 + 2xy + 2xz + 2xw + 2yz + 2yw + 2zw.$$

4.5 DIFFERENCE OF SQUARES

In this section, we continue our study of special products.

Theorem 4.6 (Difference of Squares) For all real numbers a, b the following identity holds

$$(a + b)(a - b) = a^2 - b^2.$$

Example 4.34 $(4x - 5y)(4x + 5y) = 16x^2 - 25y^2$.

Example 4.35 $(xy - 5y)(xy + 5y) = x^2y^2 - 25y^2$.

Example 4.36 Show that $(a - b - c)(a + b + c) = a^2 - b^2 - c^2 - 2bc$.

Solution: ▶ *We have*

$$(a - b - c)(a + b + c) = (a - (b + c))(a + (b + c))$$

$$= a^2 - (b + c)^2$$

$$= a^2 - (b^2 + 2bc + c^2)$$

$$= a^2 - b^2 - c^2 - 2bc,$$ ◀

Example 4.37

$$(x + 1)^2 - (x - 1)^2 = ((x + 1) + (x - 1))((x + 1) - (x - 1))$$

$$= (2x)(2)$$

$$= 4x$$

Example 4.38 We have

$$(a - 1)(a + 1)(a^2 + 1)(a^4 + 1)(a^8 + 1) = (a^2 - 1)(a^2 + 1)(a^4 + 1)(a^8 + 1)$$

$$= (a^4 - 1)(a^4 + 1)(a^8 + 1)$$

$$= (a^8 - 1)(a^8 + 1)$$

$$= a^{16} - 1.$$

Example 4.39 We have

$$(x^2 - x + 1)(x^2 + x + 1) = ((x^2 + 1) - x)((x^2 + 1) + x)$$

$$= (x^2 + 1)^2 - x^2$$

$$= x^4 + 2x^2 + 1 - x^2$$

$$= x^4 + x^2 + 1.$$

Example 4.40 Prove that

$$x^2 - (x + 2)(x - 2) = 4.$$

Solution: ▶ *We have*

$$x^2 - (x + 2)(x - 2) = x^2 - (x^2 - 4)$$

$$= x^2 - x^2 + 4$$

$$= 4$$

Explain how to find the exact value of the following, *mentally*:

(987654321) (987654321) − (987654323) (987654319),

Since the first term is two values away from the third term, we can set x = 987654321. Then x + 2 = 987654323 and x − 2 = 987654319, from where (987654321)(987654321) − (987654323)(987654319) = 4

follows. ◀

Supplementary Exercises

Exercise 4.5.1 *Multiply and collect like terms:* $(a + 4)^2 - (a + 2)^2$.

Exercise 4.5.2 *Prove that*

$$(x - y)(x + y)(x^2 + y^2) = x^4 - y^4.$$

Exercise 4.5.3 *Multiply and collect like terms:* $(a + 1)^4 - (a - 1)^4$.

Exercise 4.5.4 *Show, without expressly computing any power of* **2**, *that*

$$(2 + 1)(2^2 + 1)(2^4 + 1)(2^8 + 1)(2^{16} + 1) = 2^{32} - 1.$$

Exercise 4.5.5 *Find*

$$1^2 - 2^2 + 3^2 - 4^2 + \cdots + 99^2 - 100^2.$$

Exercise 4.5.6 Without using a calculator, *determine the exact numerical value of*

$$(123456789)^2 - (123456787)(123456791).$$

Exercise 4.5.7 Without using a calculator, *determine the exact numerical value of*

$$(666\ 666\ 666)^2 - (333\ 333\ 333)^2.$$

4.6 CUBE OF A SUM

We introduce now a special product formula for the cube of a sum. This formula is valid for both real and complex numbers.

Theorem 4.7 Let a and b be real numbers. Then

$$(a + b)^3 = a^3 + 3a^2b + 3ab^2 + b^3.$$

Corollary 4.1 Let a and b be real numbers. Then

$$(a - b)^3 = a^3 - 3a^2b + 3ab^2 - b^3.$$

Example 4.41 We have,

$$(x + 2y)^3 = x^3 + 3x^2(2y) + 3x(2y)^2 + (2y)^3 = x^3 + 6x^2y + 12xy^2 + 8y^3.$$

Example 4.42 We have,

$$(3x - 2y)^3 = (3x)^3 - 3(3x)^2(2y) + 3(3x)(2y)^2 - (2y)^3 = 27x^3 - 54x^2y + 36xy^2 - 8y^3.$$

Observe that

$$3ab(a + b) = 3a^2b + 3ab^2,$$

therefore, we may write the cube of a sum identity as

$$(a + b)^3 = a^3 + b^3 + 3ab(a + b).$$

Example 4.43 Given that $a + b = 2$ and $ab = 3$, find $a^3 + b^3$.

Solution: ▶ *We have*

$$a^3 + b^3 = (a + b)^3 - 3ab\,(a + b) = 2^3 - 3(3)\,(2) = 8 - 18 = -10.$$

◀

Supplementary Exercises

Exercise 4.6.1 *Expand and simplify:* $(5x + 1)^3$.

Exercise 4.6.2 *Multiply and collect like terms:* $(x + 1)^3 + (x + 1)^2$.

Exercise 4.6.3 *Multiply and collect like terms:* $(x - 1)^3 - (x + 2)^2$.

Exercise 4.6.4 *Multiply and collect like terms:* $(x - 2)^3 - (x - 2)^2$.

Exercise 4.6.5 *Given that* $a - b = 6$ *and* $ab = 3$, *find* $a^3 - b^3$.

Exercise 4.6.6 *Given that* $a + 2b = 6$ *and* $ab = 3$, *find* $a^3 + 8b^3$.

4.7 SUM AND DIFFERENCE OF CUBES

In this section, we study our final special product.

Theorem 4.8 (Sum and Difference of Cubes) For all real numbers a, b the following identity holds

$$(a \pm b)(a^2 \mp ab + b^2) = a^3 \pm b^3.$$

Example 4.44 $(a - 2b)(a^2 + 2ab + 4b^2) = a^3 - 8b^3$.

Example 4.45 $(5a + 2b)(25a^2 - 10ab + 4b^2) = 125a^3 - 8b^3$.

Supplementary Exercises

Exercise 4.7.1 *Expand and simplify:* $(x - 2)(x^2 + 2x + 4)$.

Exercise 4.7.2 *Expand and simplify:* $(x + 8)(x^2 - 8x + 64)$.

Exercise 4.7.3 *Given that* $a - b = 6$ *and* $ab = 3$, *find* $a^3 - b^3$.

DIVISION

This chapter reviews the mathematical operations of division in algebraic expressions. We introduce the concepts of *Term-by-Term Division, Long Division, Factoring, Special Factorizations*, and *Rational Expressions*.

5.1 TERM-BY-TERM DIVISION

From the distributive law we deduce

$$\frac{b+c}{a} = \frac{1}{a}(b+c) = \frac{b}{c} + \frac{c}{a}.$$

Example 5.1 $\dfrac{x^2 - 2xy}{x} = \dfrac{x^2}{x} - \dfrac{2xy}{x} = x - 2y.$

☞ *Be careful not to confuse an expression of the form $\frac{a+b}{c}$ with one of the form $\frac{ab}{c}$. Also note that an expression of the form $\frac{ab}{c}$ can be evaluated in various ways. For example, one may evaluate first the product ab and then divide this by c, or one may divide a by c and then multiply this by b, etc.*

Example 5.2 We have, $\dfrac{(x^2)(-2xy)}{x} = \dfrac{-2x^3 y}{x} = -2x^2 y.$

This may also be evaluated in the following manner,

$$\frac{(x^2)(-2xy)}{x} = \frac{x^2}{x}(-2xy) = x(-2xy) = -2x^2 y,$$

as before.

Example 5.3 $\dfrac{-24x^6 - 32x^4}{-8x^3} = \dfrac{-24x^6}{-8x^3} - \dfrac{32x^4}{-8x^3} - 3x^3 - (-4x) = 3x^3 + 4x.$

Example 5.4 We have,

$$\frac{(-24x^6) - (32x^4)}{-8x^3} = \frac{768x^{10}}{-8x^3} = -96x^7.$$

We may also proceed as follows,

$$\frac{(-24x^6)-(32x^4)}{-8x^3}=\frac{-24x^6}{-8x^3}(-32x^4)=(3x^3)(-32x^4)=-96x^7.$$

When a negative exponent arises in a term, we prefer to express the term with positive exponents.

Example 5.5

$$\frac{12x^2y^3+6x^4y^2}{3x^3y^3}=\frac{12x^2y^3}{3x^3y^3}+\frac{6x^4y^2}{3x^3y^3}$$
$$=4x^{-1}+2xy^{-1}$$
$$=\frac{4}{x}+\frac{2x}{y}.$$

Supplementary Exercises

Exercise 5.1.1 *Perform the division:* $\frac{q^2-pq-pqr}{-q}$.

Exercise 5.1.2 *Divide term by term:* $\frac{34x^3y^2+51x^2y^3}{17xy}$.

Exercise 5.1.3 *Divide:* $\frac{34x^3y^2+(51x^2y^3)}{17xy}$.

Exercise 5.1.4 *Divide term by term:* $\frac{6aaaaxxx+4aaaxxx}{2aaxx}$.

Exercise 5.1.5 *Divide:* $\frac{(6aaaaxxx)+(4aaaxxx)}{2aaxx}$.

Exercise 5.1.6 *Divide term by term:* $\frac{x^2-xy-xz}{-x}$.

Exercise 5.1.7 *Divide term by term:* $\frac{5a^2b-7ab^3}{-ab}$.

Exercise 5.1.8 *Divide:* $\frac{(5a^2b)(-7ab^3)}{-ab}$.

Exercise 5.1.9 *Divide:* $\frac{(x^2)(-xy)(-xz)}{-x}$.

5.2 LONG DIVISION

We must now confront the problem when the divisor consists of more than one term. The algorithm for algebraic long division resembles that for long division of natural numbers.

Example 5.6 Find $(-x+x^2-6)\div(x+2)$.

Solution: ▶ *First, we rewrite* $-x+x^2-6$ *so that the exponents of each of the terms are listed in decreasing order:* x^2-x-6. *Next, display the dividend and divisor as follows:*

$$x+2\overline{)x^2-x-6}.$$

Now by what must x (the term with the largest degree in the divisor) be multiplied to return x^2 (the term with the largest degree in the dividend)? The answer is $\frac{x^2}{x} = x$, and so we write x in the quotient,

$$x + 2 \overline{)x^2 - x - 6} \quad \overset{x}{} \; .$$

Multiply this x of the quotient by the divisor, obtaining $x(x + 2) = x^2 + 2x$ and change all signs:

$$\begin{array}{r} x \\ x + 2 \overline{)x^2 - x - 6} \\ \underline{-x^2 - 2x} \end{array}$$

Add, and obtain

$$\begin{array}{r} x \\ x + 2 \overline{)x^2 - x - 6} \\ \underline{-x^2 - 2x} \\ -3x - 6 \end{array}$$

The new dividend is now $-3x - 6$. Again by what must x (the term with the largest degree in the divisor) be multiplied to return $-3x$ (the term with the largest degree in the dividend)? The answer is $\frac{-3x}{x} = -3$, and so we write -3 in the quotient,

$$\begin{array}{r} x - 3 \\ x + 2 \overline{)x^2 - x - 6} \\ \underline{-x^2 - 2x} \\ -3x - 6 \end{array}$$

Multiply this -3 of the quotient by the divisor, obtaining $-3(x + 2) = -3x - 6$ and change all signs:

$$\begin{array}{r} x - 3 \\ x + 2 \overline{)x^2 - x - 6} \\ \underline{-x^2 - 2x} \\ -3x - 6 \\ \underline{3x + 6} \end{array}$$

Add, and obtain

$$\begin{array}{r} x - 3 \\ x + 2 \overline{)x^2 - x - 6} \\ \underline{-x^2 - 2x} \\ -3x - 6 \\ \underline{3x + 6} \\ 0 \end{array}$$

Since the remainder is **0**, *the division ends and we conclude that*

$$(-x + x^2 - 6) \div (x + 2) = \frac{x^2 - x - 6}{x + 2} = x + 3.$$ ◀

Example 5.7 Find $(3x^4 - 2x^3 - 9x^2 + 5x - 6) \div (x - 2)$.

Solution: ▶ *The terms of* $3x^4 - 2x^3 - 9x^2 + 5x - 6$ *are already written with exponents in decreasing order. Next, display the dividend and divisor as follows:*

$$x - 2 \overline{) 3x^4 - 2x^3 - 9x^2 + 5x - 6}.$$

Now, what must x *(the term with the largest degree in the divisor) be multiplied by to return the expression* $3x^4$ *(the term with the largest degree in the dividend)? The answer is* $\frac{3x^4}{x} = 3x^3$, *and so we write* x *in the quotient,*

$$
\begin{array}{r}
3X^3 \\
x - 2 \overline{) 3x^4 - 2x^3 - 9x^2 - 5x - 6} \,.
\end{array}
$$

Multiply this $3x^3$ *of the quotient by the divisor, obtaining* $3x^3(x - 2) = 3x^4 - 6x^3$ *and change all signs:*

$$
\begin{array}{r}
3x^3 \\
x - 2 \overline{) 3x^4 - 2x^3 - 9x^2 + 5x - 6} \\
\underline{-3x^4 + 6x^3}
\end{array}
$$

Add, and obtain

$$
\begin{array}{r}
3x^3 \\
x - 2 \overline{) 3x^4 - 2x^3 - 9x^2 + 5x - 6} \\
\underline{-3x^4 + 6x^3} \\
4x^3 - 9x^2
\end{array}
$$

The new dividend is now $4x^3 - 9x^2$. *Again by what must* x *(the term with the largest degree in the divisor) be multiplied to return* $4x^3$ *(the term with the largest degree in the dividend)? The answer is* $\frac{4x^3}{x} = 4x^2$, *and so we write* $+4x^2$ *in the quotient,*

$$
\begin{array}{r}
3x^3 + 4x^2 \\
x - 2 \overline{) 3x^4 - 2x^3 - 9x^2 + 5x - 6} \\
\underline{-3x^4 + 6x^3} \\
4x^3 - 9x^2
\end{array}
$$

Multiply this $+4x^2$ *of the quotient by the divisor, obtaining* $+4x^2(x - 2) = 4x^3 - 8x^2$ *and change all signs:*

$$
\begin{array}{r}
3x^3 + 4x^2 \\
x - 2 \overline{) 3x^4 - 2x^3 - 9x^2 + 5x - 6} \\
\underline{-3x^4 + 6x^3} \\
4x^3 - 9x^2 \\
\underline{-4x^3 + 8x^2}
\end{array}
$$

Add, and obtain

$$
\begin{array}{r}
3x^3 + 4x^2 \\
x-2\overline{)3x^4 - 2x^3 - 9x^2 + 5x - 6} \\
\underline{-3x^4 + 6x^3} \\
4x^3 - 9x^2 \\
\underline{-4x^3 + 8x^2} \\
-x^2 + 5x
\end{array}
$$

The new dividend is now $-x^2 + 5x$. Again by what must x (the term with the largest degree in the divisor) be multiplied to give $-x^2$ (the term with the largest degree in the dividend)? The answer is $\frac{-x^2}{x} = -x$, and so we write $-x$ in the quotient,

$$
\begin{array}{r}
3x^3 + 4x^2 - x \\
x-2\overline{)3x^4 - 2x^3 - 9x^2 + 5x - 6} \\
\underline{-3x^4 + 6x^3} \\
4x^3 - 9x^2 \\
\underline{-4x^3 + 8x^2} \\
-x^2 + 5x
\end{array}
$$

Multiply this $-x$ of the quotient by the divisor, obtaining $-x(x-2) = -x^2 + 2x$ and change all signs:

$$
\begin{array}{r}
3x^3 + 4x^2 - x \\
x-2\overline{)3x^4 - 2x^3 - 9x^2 + 5x - 6} \\
\underline{-3x^4 + 6x^3} \\
4x^3 - 9x^2 \\
\underline{-4x^3 + 8x^2} \\
-x^2 + 5x \\
\underline{x^2 - 2x}
\end{array}
$$

Add, and obtain

$$
\begin{array}{r}
3x^3 + 4x^2 - x \\
x-2\overline{)3x^4 - 2x^3 - 9x^2 + 5x - 6} \\
\underline{-3x^4 + 6x^3} \\
4x^3 - 9x^2 \\
\underline{-4x^3 + 8x^2} \\
-x^2 + 5x \\
\underline{x^2 - 2x} \\
3x - 6
\end{array}
$$

The new dividend is now $3x - 6$. Again by what must x (the term with the largest degree in the divisor) be multiplied to give $3x$ (the term with the largest degree in the dividend)? The answer is $\frac{3x}{x} = 3$, and so we write 3 in the quotient,

$$
\begin{array}{r}
3x^3 + 4x^2 - x + 3 \\
x - 2 \overline{)\,3x^4 - 2x^3 - 9x^2 + 5x - 6} \\
\underline{-3x^4 + 6x^3} \\
4x^3 - 9x^2 \\
\underline{-4x^3 + 8x^2} \\
-x^2 + 5x \\
\underline{x^2 - 2x} \\
3x - 6
\end{array}
$$

Multiply this **3** *of the quotient by the divisor, obtaining* $3(x - 2) = 3x - 6$ *and change all signs:*

$$
\begin{array}{r}
3x^3 + 4x^2 - x + 3 \\
x - 2 \overline{)\,3x^4 - 2x^3 - 9x^2 + 5x - 6} \\
\underline{-3x^4 + 6x^3} \\
4x^3 - 9x^2 \\
\underline{-4x^3 + 8x^2} \\
-x^2 + 5x \\
\underline{x^2 - 2x} \\
3x - 6 \\
\underline{-3x + 6}
\end{array}
$$

Add, and obtain

$$
\begin{array}{r}
3x^3 + 4x^2 - x + 3 \\
x - 2 \overline{)\,3x^4 - 2x^3 - 9x^2 + 5x - 6} \\
\underline{-3x^4 + 6x^3} \\
4x^3 - 9x^2 \\
\underline{-4x^3 + 8x^2} \\
-x^2 + 5x \\
\underline{x^2 - 2x} \\
3x - 6 \\
\underline{-3x + 6} \\
0
\end{array}
$$

Since the remainder is **0**, *the division ends and we conclude that*

$$
(3x^4 - 2x^3 - 9x^2 + 5x - 6) \div (x - 2) = \frac{3x^4 - 2x^3 - 9x^2 + 5x - 6}{x - 2} = 3x^3 + 4x^2 - x + 3.
$$ ◀

We now give an example where there is a remainder.

Example 5.8 Find $(2x^2 + 1) \div (x + 1)$.

Solution: ▶ *Display the dividend and divisor as follows:*

$$x+1\overline{)2x^2 \quad +1}\,.$$

Now by what must x (the term with the largest degree in the divisor) be multiplied to give $2x^2$ (the term with the largest degree in the dividen)d? The answer is $\frac{2x^2}{x} = 2x$, and so we write $2x$ in the quotient,

$$x+1\overline{)2x^2 \quad +1}^{\,2x}\,.$$

Multiply this $2x$ of the quotient by the divisor, obtaining $2x(x + 1) = 2x^2 + 2x$ and change all signs:

$$\begin{array}{r} 2x \\ x+1\overline{)2x^2 \quad +1} \\ \underline{-2x^2 - 2x} \end{array}$$

Add, and obtain

$$\begin{array}{r} 2x \\ x+1\overline{)2x^2 \quad +1} \\ \underline{-2x^2 - 2x} \\ -2x+1 \end{array}$$

The new dividend is now $-2x + 1$. Again by what must x (the term with the largest degree in the divisor) be multiplied to give $-2x$ (the term with the largest degree in the dividend)? The answer is $\frac{-2x}{x} = -2$, and so we write -2 in the quotient,

$$\begin{array}{r} 2x - 2 \\ x+1\overline{)2x^2 \quad +1} \\ \underline{-2x^2 - 2x} \\ -2x+1 \end{array}$$

Multiply this -2 of the quotient by the divisor, obtaining $-2(x + 1) = -2x - 2$ and change all signs:

$$\begin{array}{r} 2x - 2 \\ x+1\overline{)2x^2 \quad +1} \\ \underline{-2x^2 - 2x} \\ -2x+1 \\ \underline{2x+2} \end{array}$$

Add, and obtain

$$
\begin{array}{r}
2x-2 \\
x+1\overline{)2x^2+1} \\
\underline{-2x^2-2x} \\
-2x+1 \\
\underline{2x+2} \\
3
\end{array}
$$

*Now, the remainder is **3**, which has lower degree than the divisor **x + 1**, therefore the division ends. We write*

$$(2x^2+1)\div(x+1)=\frac{2x^2+1}{x+1}=2x-2+\frac{3}{x+1}. \qquad \blacktriangleleft$$

We provide below additional examples, without explanations.

Example 5.9 We have,

$$
\begin{array}{r}
x^2-3x+9 \\
x+3\overline{)x^3+27} \\
\underline{-x^3-3x^2} \\
-3x^2 \\
\underline{3x^2+9x} \\
9x+27 \\
\underline{-9x-27} \\
0
\end{array}
$$

Example 5.10 We have,

$$
\begin{array}{r}
x^3-2x^2+4x+8 \\
x+2\overline{)x^4-16} \\
\underline{-x^4+2x^3} \\
2x^3 \\
\underline{-2x^3+4x^2} \\
4x^2 \\
\underline{-4x^2+8x} \\
8x-16 \\
\underline{-8x+16} \\
0
\end{array}
$$

Example 5.11 We have,

$$
\begin{array}{r}
x^2 + 4x + 4 \\
x^2 - x - 2 \overline{\smash{\big)}\ x^4 + 3x^3 - 2x^2 + x - 1} \\
\underline{-x^4 + x^3 + 2x^2} \\
4x^3 \qquad\quad + x \\
\underline{-4x^3 + 4x^2 + 8x} \\
4x^2 + 9x - 1 \\
\underline{-4x^2 + 4x + 8} \\
13x + 7
\end{array}
$$

and therefore we write

$$
\frac{x^4 + 3x^3 - 2x^2 + x - 1}{x^2 - x - 2} = x^2 + 4x + 4 + \frac{13x + 7}{x^2 - x - 2}.
$$

Supplementary Exercises

Exercise 5.2.1 *Perform the division:* $(x^2 - 2x + 1) \div (x - 1)$.

Exercise 5.2.2 *Perform the division:* $(x^2 + 5x + 6) \div (x + 2)$.

Exercise 5.2.3 *Perform the division:* $(x^3 + 1) \div (x + 1)$.

Exercise 5.2.4 *Expand and collect like terms:*

$$
\frac{x^3 - 8}{x - 2} - x^2.
$$

Exercise 5.2.5 *Expand and collect like terms:*

$$
\frac{6x^3 - x^2 - 4x - 1}{2x + 1} - \frac{3x^3 + 3x - x^2 - 1}{3x - 1}.
$$

5.3 FACTORING I

Definition 5.1 To *factor* an algebraic expression is to express it as a product.

The idea behind factoring is essentially that of using the distributive law backwards,

$$ab + ac = a(b + c),$$

where the dextral quantity is a decomposition into factors of the sinistral quantity.

We will begin by giving examples where the greatest common divisor of the terms is different from **1**, and therefore it may be removed. Thus an expression with two terms will be factored as

$$a + b = \blacksquare(\Box_1 + \Box_2),$$

an expression with three terms will be factored as

$$a + b + c = \blacksquare\ (\Box_1 + \Box_2 + \Box_3),$$

and so on.

Example 5.12 Factor $20a^2b^3 + 24a^3b^2$.

Solution: ▶ *The greatest common divisor of **20** and **24** is **4**, therefore it can be removed. Also for the two expressions a^2 and a^3, their greatest common divisor is the one with the least exponent, that is a^2. Similarly, the greatest common divisor of b^2 and b^3 is b^2. Thus $4a^2b^2$ is a common factor of both terms and the desired factorization has the form*

$$20a^2b^3 + 24a^3b^2 = 4a^2b^2(\square_1 + \square_2).$$

We need to determine $\square_1$ and $\square_2$. We have

$$20a^2b^3 = 4a^2b^2\square_1 \Rightarrow \square_1 = 5b,$$

and

The desired factorization is thus

$$20a^2b^3 + 24a^3b^2 = 4a^2b^2(5b + 6a). \qquad \blacktriangleleft$$

☞ *Multiplying, $4a^2b^2(5b + 6a) = 20a^2b^3 + 24a^3b^2$.*

Example 5.13 Factor $-30x^2yza^2b + 36xy^2za + 16x^3y^2z^2b$.

Solution: ▶ *The greatest common divisor of $-30, 36, 16$ is **2**. The letters x, y, z appear in all three terms, but a and b do not appear in all three terms. The least power of x appearing in all three terms is **1**, and same for y and z. Thus the greatest common divisor of the three terms is $2xyz$ and the desired factorization is*

$$2xyz(-15xa^2b + 18ya + 8x^2yzb). \qquad \blacktriangleleft$$

Example 5.14 Factor $ab + a^2b + ab^2$.

Solution: ▶ *The desired factorization is clearly*

$$ab + a^2b + ab^2 = ab(1 + a + b). \qquad \blacktriangleleft$$

Example 5.15 Prove that the sum of two even integers is even.

Solution: ▶ *Let $2a$ and $2b$ be two even integers. Then*

$$2a + 2b = 2(a + b),$$

that is, twice the integer $a + b$, and therefore an even integer. $\blacktriangleleft$

Example 5.16 Factor $14a^3b^2(x + y)^2(x - y)^3 + 20ab^3(x + y)^3(x - y)^4$.

Solution: ▶ *The desired factorization is clearly*

$$14a^3b^2(x+y)^2(x-y)^3 + 20ab^3(x+y)^3(x-y)^4 = 2ab^2(x+y)^2(x-y)^3(7a^2 + 10b(x+y)(x-y)). \qquad \blacktriangleleft$$

Example 5.17 Resolve into factors: $ax^2 + a + bx^2 + b$.

Solution: ▶ *In this case, no factor greater than **1** is common to all terms. We have, however,*

$$ax^2 + a + bx^2 + b = a(x^2 + 1) + b(x^2 + 1).$$

Notice now that $x^2 + 1$ is a common factor of both summands, and finally,

$$ax^2 + a + bx^2 + b = a(x^2 + 1) + b(x^2 + 1) = (x^2 + 1)(a + b). \quad \blacktriangleleft$$

Example 5.18 Resolve into factors: $x^3 + x^2 + x + 1$.

Solution: ▶ *Like in the preceding problem, we have no factor greater than 1 being common to all terms. Observe that*

$$x^3 + x^2 + x + 1 = x^2(x + 1) + 1(x + 1).$$

Notice now that $x + 1$ is a common factor of both summands, and finally,

$$x^3 + x^2 + x + 1 = x^2(x + 1) + 1(x + 1) = (x + 1)(x^2 + 1).$$

Example 5.19 Resolve into factors: $x^2 - ax + bx - ab$.

Solution: ▶ *We have,*

$$x^2 - ax + bx - ab = x(x - a) + b(x - a) = (x - a)(x + b). \quad \blacktriangleleft$$

Sometimes it is necessary to rearrange the summands.

Example 5.20 Resolve into factors: $12a^2 - 4ab - 3ax^2 + bx^2$.

Solution: ▶ *We have,*

$$12a^2 - 4ab - 3ax^2 + bx^2 = (12a^2 - 3ax^2) - (4ab - bx^2) = 3a(4a - x^2) - b(4a - x^2)$$

$$= (4a - x^2)(3a - b). \quad \blacktriangleleft$$

Supplementary Exercises

Exercise 5.3.1 *Factor $-2a^2b^3$ from $-4a^6b^5 + 6a^3b^8 - 12a^5b^3 - 2a^2b^3$.*

Exercise 5.3.2 *Factor $2a^3b^3$ from $4a^3b^4 - 10a^4b^3$.*

Exercise 5.3.3 *Factor $\frac{3}{4}x$ from $\frac{9}{16}x^2 - \frac{3}{4}x$.*

Exercise 5.3.4 *Factor -1 from $-x + 2y - 3z$.*

Exercise 5.3.5 *Prove that the sum of two odd integers is even.*

Exercise 5.3.6 *Resolve into factors: $x^3 - x^2$.*

Exercise 5.3.7 *Factor $125a^4b^5c^5 - 45a^5b^3c^4 + 5a^3b^2c^4 - 300a^4b^2c^8 - 10a^3b^2c^5$.*

Exercise 5.3.8 *Resolve into factors: $5x^5 - 10a^7x^3 - 15a^3x^3$.*

Exercise 5.3.9 *Factor $3x^3y + 4x^2y^3 - 6x^6 - 10x^4$.*

Exercise 5.3.10 *Factor $3m^6p^4q^2 - 9m^5p^2qx + 3m^7p^3qx + 3m^4p^2q - 6m^5p^4qx^2y$.*

Exercise 5.3.11 *Resolve into factors: $38a^2x^5 + 57a^4x^2$.*

Exercise 5.3.12 *Decompose into factors: $a^2 + ab + ac + bc$.*

Exercise 5.3.13 *Decompose into factors: $a^2 - ab + ac - bc$.*

Exercise 5.3.14 *Decompose into factors: $y^3 - y^2 + y - 1$.*

Exercise 5.3.15 *Decompose into factors: $2x^3 + 3x^2 + 2x + 3$.*

Exercise 5.3.16 *Decompose into factors: $a^2x + abx + ac + aby + b^2y + bc$.*

5.4 FACTORING II

In this section, we study factorizations of expressions of the form $ax^2 + bx + c$, the so-called *quadratic trinomials.* Let us start with expressions of the form $x^2 + bx + c$. Suppose that

$$x^2 + bx + c = (x + p)(x + q).$$

Then, upon multiplying,

$$x^2 + bx + c = (x + p)(x + q) = x^2 + (p + q)x + pq.$$

Thus $c = pq$ and $b = p + q$, that is, the constant term is the product of two numbers, and the coefficient of x is the sum of these two numbers.

Example 5.21 To factor $x^2 + 5x + 6$, we look for two numbers whose product is **6** and whose sum is **5**. Clearly **2** and **3** are these two numbers and so,

$$x^2 + 5x + 6 = (x + 2)(x + 3).$$

You should multiply the last product to verify the equality.

Example 5.22 To factor $x^2 + 7x + 6$, we look for two numbers whose product is **6** and whose sum is **7**. Clearly **1** and **6** are these two numbers and so,

$$x^2 + 7x + 6 = (x + 1)(x + 6).$$

You should multiply the last product to verify the equality.

Example 5.23 To factor $x^2 - 5x + 6$, we look for two numbers whose product is **6** and whose sum is **−5**. Clearly **−2** and **−3** are these two numbers and so,

$$x^2 - 5x + 6 = (x - 2)(x - 3).$$

Example 5.24 To factor $x^2 - x - 6$, we look for two numbers whose product is **−6** and whose sum is **−1**. Clearly **−3** and **+2** are these two numbers and so,

$$x^2 - x - 6 = (x - 3)(x + 2).$$

Example 5.25 To factor $x^2 + x - 6$, we look for two numbers whose product is **−6** and whose sum is **+1**. Clearly **+3** and **−2** are these two numbers and so,

$$x^2 + x - 6 = (x + 3)(x - 2).$$

Example 5.26 Following are additional examples to inspect:

$$x^2 + 10x + 24 = (x + 4)(x + 6),$$

$$x^2 + 11x + 24 = (x + 3)(x + 8),$$

$$x^2 + 14x + 24 = (x + 2)(x + 12),$$

$$x^2 + 25x + 24 = (x + 1)(x + 24).$$

Example 5.27 Following are additional examples to inspect:

$$x^2 - 10x + 24 = (x - 4)(x - 6),$$

$$x^2 - 11x + 24 = (x - 3)(x - 8),$$

$$x^2 - 14x + 24 = (x - 2)(x - 12),$$

$$x^2 - 25x + 24 = (x - 1)(x - 24).$$

Example 5.28 Following are additional examples to inspect:

$$x^2 + 2x - 24 = (x - 4)(x + 6),$$

$$x^2 - 5x - 24 = (x + 3)(x - 8),$$

$$x^2 + 10x - 24 = (x - 2)(x + 12),$$

$$x^2 - 10x - 24 = (x + 2)(x - 12).$$

In an instance when the coefficient of the x^2 is different from **1**, more cases must be considered. The exercise is more difficult, but the method is essentially the same.

Example 5.29 By inspection,

$$5x^2 + 22x + 8 = (5x + 2)(x + 4),$$

$$5x^2 - 22x + 8 = (5x - 2)(x - 4),$$

$$5x^2 - 18x - 8 = (5x + 2)(x - 4),$$

$$5x^2 + 18x - 8 = (5x - 2)(x + 4),$$

$$5x^2 + 39x - 8 = (5x - 1)(x + 8).$$

Example 5.30 By inspection,

$$20x^2 + 43x + 21 = (5x + 7)(4x + 3),$$

$$20x^2 + 47x + 21 = (5x + 3)(4x + 7),$$

$$20x^2 + 76x + 21 = (10x + 3)(2x + 7),$$

$$20x^2 + 143x + 21 = (20x + 3)(x + 7).$$

The inquiring reader may wonder whether all quadratic trinomials $ax^2 + bx + c$ factor in the manner above, that is, into linear factors for which all coefficients are integers. The answer is *no*. For example, if $x^2 - 2$ does not factor into linear factors whose coefficients are integers because there are no two integers whose product is **−2** and whose sum is **0**. If we augment our choice for the coefficients for linear factors, then we may write

$$x^2 - 2 = (x - \sqrt{2})(x + \sqrt{2}),$$

but we will not consider these cases in this book.

Supplementary Exercises

Exercise 5.4.1 *Decompose into factors:* $a^2 - 11a + 30$.

Exercise 5.4.2 *Decompose into factors:* $a^2 - 38a + 361$.

Exercise 5.4.3 *Decompose into factors:* $a^4b^4 + 37a^2b^2 + 300$.

Exercise 5.4.4 *Decompose into factors:* $x^2 - 23x + 132$.

Exercise 5.4.5 *Decompose into factors:* $x^4 - 29x^2 + 204$.

Exercise 5.4.6 *Decompose into factors:* $x^2 + 35x + 216$.

Exercise 5.4.7 *Resolve into factors:* $5x^2 + 17x + 6$.

Exercise 5.4.8 *Resolve into factors:* $14x^2 + 29x - 15$.

5.5 SPECIAL FACTORIZATIONS

Recall the special products we studied in the Chapter 4 on multiplcation. We list them here for easy reference.

Difference of Squares

$$x^2 - y^2 = (x - y)(x + y).$$

Difference of Cubes

$$x^3 - y^3 = (x - y)(x^2 + xy + y^2).$$

Sum of Cubes

$$x^3 + y^3 = (x + y)(x^2 - xy + y^2).$$

Square of a Sum

$$x^2 + 2xy + y^2 = (x + y)^2.$$

Square of a Difference

$$x^2 - 2xy + y^2 = (x - y)^2.$$

We will use these special factorizations and the methods of the preceding sections to treat more complicated factorization problems.

Example 5.31 To factor $x^4 - 9y^2$, observe that it is a difference of squares, and so

$$x^4 - 9y^2 = (x^2 - 3y)(x^2 + 3y).$$

Example 5.32 To factor $x^3 - 1$, observe that it is a difference of cubes, and so

$$x^3 - 1 = (x - 1)(x^2 + x + 1).$$

Sometimes we need to use more than one method.

Example 5.33 To factor $x^3 - 4x$, factor a common factor and then the difference of squares:

$$x^3 - 4x = x(x^2 - 4) = x(x - 2)(x + 2).$$

Example 5.34 To factor $x^4 - 81$, observe that there are two difference of squares:

$$x^4 - 81 = (x^2 - 9)(x^2 + 9) = (x - 3)(x + 3)(x^2 + 9).$$

Example 5.35 To factor $x^6 - 1$, observe that there is a difference of squares and then a sum and a difference of cubes:

$$x^6 - 1 = (x^3 - 1)(x^3 + 1) = (x - 1)(x^2 + x + 1)(x + 1)(x^2 - x + 1).$$

The following method is useful to convert some expressions into differences of squares.

Example 5.36 We have

$$x^4 + x^2 + 1 = x^4 + 2x^2 + 1 - x^2$$
$$= (x^2 + 1)^2 - x^2$$
$$= (x^2 + 1 - x)(x^2 + 1 + x).$$

Example 5.37 We have

$$x^4 + 4 = x^4 + 4x^2 + 4 - 4x^2$$
$$= (x^2 + 2)^2 - 4x^2$$
$$= (x^2 + 2 - 2x)(x^2 + 2 + 2x).$$

This trick is often used in factoring quadratic trinomials, where it is often referred to as the technique of *completing the square*. We will give some examples of factorizations that may also be obtained using the trial– and -error method of the preceding section.

Example 5.38 We have

$$x^2 - 8x - 9 = x^2 - 8x + 16 - 9 - 16$$
$$= (x - 4)^2 - 25$$
$$= (x - 4)^2 - 5^2$$
$$= (x - 4 - 5)(x - 4 + 5)$$
$$= (x - 9)(x + 1).$$

Example 5.39 We have

$$x^2 + 4x - 117 = x^2 + 4x + 4 - 117 - 4$$
$$= (x + 2)^2 - 11^2$$
$$= (x + 2 - 11)(x + 2 + 11)$$
$$= (x - 9)(x + 13).$$

The techniques learned may be used to solve some purely arithmetic problems.

Example 5.40 Find the prime factor greater than **9000** of

$$1, 000, 002, 000, 001.$$

Solution: ▶ *We have*

$$1, 000, 002, 000, 001 = 10^{12} + 2 \cdot 10^6 + 1$$
$$= (10^6 + 1)^2$$
$$= ((10^2)^3 + 1)^2$$
$$= (10^2 + 1)^2 ((10^2)^2 - 10^2 + 1)^2$$
$$= 101^2 9901^2,$$

Therefore, the prime factor greater than 9000 is **9901**. ◀

Supplementary Exercises

> **Exercise 5.5.1** *Given that $x + 2y = 3$ and $x - 2y = -1$, find $x^2 - 4y^2$.*
>
> **Exercise 5.5.2** *Resolve into factors: $x^4 - 16$.*
>
> **Exercise 5.5.3** *Resolve into factors: $(a + b)^2 - c^2$.*
>
> **Exercise 5.5.4** *Resolve into factors: $x^3 - x$.*
>
> **Exercise 5.5.5** *Find all positive primes of the form $n^3 - 8$, where n is a positive integer.*

5.6 RATIONAL EXPRESSIONS

To reduce an algebraic fraction, we must first factor it.

Example 5.41 Reduce the fraction $\dfrac{x^2 - 1}{x^2 - 2x + 1}$.

Solution: ▶ *We have* $\dfrac{x^2 - 1}{x^2 - 2x + 1} = \dfrac{(x-1)(x+1)}{(x-1)(x-1)} = \dfrac{x+1}{x-1}$. ◀

Example 5.42 Express the fraction $\dfrac{x^2 + 5x + 6}{x^2 - 5x - 14}$ in lowest terms.

Solution: ▶ *We have* $\dfrac{x^2 + 5x + 6}{x^2 - 5x - 14} = \dfrac{(x+2)(x+3)}{(x+2)(x-7)} = \dfrac{(x+3)}{(x-7)}$.

To add or subtract algebraic fractions, we first write them in common denominators. We will prefer to expand all products and collect like terms. ◀

Example 5.43 Add:

$$\frac{3}{x-3} + \frac{2}{x+2}.$$

Solution: ▶ *We have,*

$$\frac{3}{x-3} + \frac{2}{x+2} = \frac{3(x+2)}{(x-3)(x+2)} + \frac{2(x-3)}{(x+2)(x-3)}$$

$$= \frac{3x+6}{(x-3)(x+2)} + \frac{2x-6}{(x+2)(x-3)}$$

$$= \frac{5x}{(x-3)(x+2)}$$

$$= \frac{5x}{x^2 - x - 6}.$$ ◀

Example 5.44 Add:

$$\frac{x}{x^2 + 2x + 1} + \frac{x}{x^2 - 1}.$$

Solution: ▶ *We have,*

$$\frac{x}{x^2+2x+1}+\frac{x}{x^2-1}=\frac{x}{(x+1)^2}+\frac{x}{(x-1)(x+1)}$$
$$=\frac{x(x-1)}{(x+1)^2(x-1)}+\frac{x(x+1)}{(x-1)(x+1)^2}$$
$$=\frac{x^2-x+x^2+x}{(x+1)^2(x-1)}$$
$$=\frac{2x^2}{x^3+x^2-x-1}$$

◀

Example 5.45 Gather all the fractions:

$$\frac{a}{b}+\frac{c}{d}-\frac{e}{f}.$$

Solution: ▶ *We have,*

$$\frac{a}{b}+\frac{c}{d}-\frac{e}{f}=\frac{adf}{bdf}+\frac{cbf}{dbf}-\frac{ebd}{fbd}$$
$$=\frac{adf+cbf-ebd}{fbd}.$$

◀

Multiplication and division of algebraic fractions is carried out in a manner similar to the operations of arithmetical fractions.

Example 5.46 Multiply:

$$\frac{x-1}{x+1}\cdot\frac{x+2}{x-2}.$$

Solution: ▶ *We have,*

$$\frac{x-1}{x+1}\cdot\frac{x+2}{x-2}=\frac{(x-1)(x+2)}{(x+1)(x-2)}$$
$$=\frac{x^2+x-2}{x^2-x-2}.$$

◀

Example 5.47 Divide:

$$\frac{x-1}{x+1}\div\frac{x+2}{x-2}.$$

Solution: ▶ *We have,*

$$\frac{x-1}{x+1} \div \frac{x+2}{x-2} = \frac{x-1}{x+1} \cdot \frac{x-2}{x+2}$$
$$= \frac{(x-1)(x-2)}{(x+1)(x+2)}$$
$$= \frac{x^2-3x+2}{x^2+3x+2}.$$ ◀

Example 5.48 The sum of two numbers is **7** and their product **21.** What is the sum of their reciprocals?

Solution: ▶ *Let x, y be the numbers. One has $x+y=7$, $xy=21$, therefore*

$$\frac{1}{x}+\frac{1}{y} = \frac{y+x}{xy} = \frac{7}{21} = \frac{1}{3}.$$ ◀

Example 5.49 Prove that

$$\frac{1}{1+\dfrac{1}{x+\dfrac{1}{x}}} = \frac{x^2+1}{x^2+x+1}.$$

Solution: ▶ *Proceeding from the innermost fraction*

$$\frac{1}{1+\dfrac{1}{x+\dfrac{1}{x}}} = \frac{1}{1+\dfrac{1}{\dfrac{x^2}{x}+\dfrac{1}{x}}}$$
$$= \frac{1}{1+\dfrac{1}{\dfrac{x^2+1}{x}}}$$
$$= \frac{1}{1+\dfrac{x}{x^2+1}}$$
$$= \frac{1}{\dfrac{x^2+1}{x^2+1}+\dfrac{x}{x^2+1}}$$
$$= \frac{1}{\dfrac{x^2+x+1}{x^2+1}}$$
$$= \frac{x^2+1}{x^2+x+1}.$$ ◀

Supplementary Exercises

Exercise 5.6.1 *Add:* $\dfrac{1}{x-1}+\dfrac{1}{x+1}$.

Exercise 5.6.2 *Subtract:* $\dfrac{1}{x-1}-\dfrac{1}{x+1}$.

Exercise 5.6.3 *Add:* $\dfrac{x}{x-2}+\dfrac{2}{x+2}$.

Exercise 5.6.4 *Subtract:* $\dfrac{x}{x-2}-\dfrac{2}{x+2}$.

Exercise 5.6.5 *Gather the fractions:* $\dfrac{1}{x}+\dfrac{1}{x-1}-\dfrac{2}{x+1}$.

Exercise 5.6.6 *Gather the fractions:* $\dfrac{3x}{2}+\dfrac{3x}{2a}-\dfrac{4x}{8a^{2}}$.

Exercise 5.6.7 *Gather the fractions:* $\dfrac{1}{s-1}-\dfrac{s}{(s-1)(s+1)}$.

Exercise 5.6.8 *Gather the fractions:* $\dfrac{x^{2}-xy}{x^{2}y}-\dfrac{y-z}{yz}-\dfrac{2z^{2}-xz}{z^{2}x}$.

Exercise 5.6.9 *Gather the fractions:* $\dfrac{2s}{s+2}-\dfrac{3}{s-2}$.

Exercise 5.6.10 *Gather the fractions:* $\dfrac{2x}{a}+\dfrac{3x}{a^{2}}-\dfrac{4x}{a^{2}}$.

Exercise 5.6.11 *The sum of two numbers is **7** and their product **21**. What is the sum of the squares of their reciprocals?*

EQUATIONS

LINEAR EQUATIONS IN ONE-VARIABLE EXPRESSIONS

This chapter applies the mathematical operations of addition, subtraction, multiplication, and division in algebraic linear equations. We introduce linear equations in sections on *Simple Equations*, *Miscellaneous Linear Equations*, and *Word Problems*.

6.1 SIMPLE EQUATIONS

Definition 6.1 An *identity* is an assertion that is true for all values of the variables involved.

Example 6.1 From the square of the sum formula we get $(x + 1)^2 = x^2 + 2x + 1$. Since this is valid for all real values of x, this is an identity.

Example 6.2 The assertion $4 + x = 6 - x$ is not an identity. For example, if $x = 0$ we have $4 + 0 = 6 - 0$, which is patently false. For $x = 1$ we have, however, $4 + 1 = 6 - 1$, which is true. Thus the assertion is sometimes true and sometimes false.

Example 6.3 The assertion $x = x + 1$ is never true. How could a real number be equal to itself plus 1?

Definition 6.2 An *equation of condition* (or *equation*, for short), is an assertion that is true for only particular values of the variables employed.

Example 6.2 is an example of an equation.

Definition 6.3 The variable whose value must be found is called the *unknown variable*. The process of finding a value that satisfies the equation is called *solving the equation*.

In this section our main focus will be linear equations with one unknown variable.

Definition 6.4 A *linear equation* in the unknown x is an equation of the form $ax + b = c$, where $a \neq 0$, and b, c are real numbers.

In order to solve simple equations, we will consider the following axioms:

Axiom 6.1 If equals are added to equals, the sums are equal.

Axiom 6.2 If equals are multiplied by equals, the products are equal.

In the following examples we will use the symbol $\Rightarrow$, which is read "implies."

Example 6.4 Solve for x:

$$x - 3 = -9$$

Solution: ▶

$$x - 3 = -9 \Rightarrow x = -9 + 3$$

$$\Rightarrow x = -6$$

Verification: $x - 3 = -6 - 3 \overset{\checkmark}{=} -9$ ◀

Example 6.5 Solve for x:

$$x + 3 = -9$$

Solution: ▶

$$x + 3 = -9 \Rightarrow x = -9 - 3$$

$$\Rightarrow x = -12.$$

Verification: $x + 3 = -12 + 3 \overset{\checkmark}{=} -9$. ◀

Example 6.6 Solve for x:

$$x + 3a = -9a$$

Solution: ▶

$$x + 3a = -9a \Rightarrow x = -3a - 9a$$

$$\Rightarrow x = -12a.$$

Verification: $x + 3a = -12a + 3a \overset{\checkmark}{=} -9a$.

Example 6.7 Solve for x:

$$x - a + b = 2a + 3b$$

Solution: ▶

$$x - a + b = 2a + 3b \Rightarrow x = 2a + 3b + a - b$$

$$\Rightarrow x = 3a + 2b.$$

Verification: $x - a + b = 3a + 2b - a + b \overset{\checkmark}{=} 2a + 3b$ ◀

Example 6.8 Solve for x:

$$2x = -88$$

Solution: ▶

$$2x = -88 \Rightarrow x = \frac{-88}{2}$$

$$\Rightarrow x = -44.$$

◀

Verification: $2x = 2(-44) \overset{\checkmark}{=} -88$. ◀

Example 6.9 Solve for x:

$$2ax = 4a$$

Solution: ▶

$$2ax = 4a \Rightarrow x = \frac{4a}{2a}$$
$$\Rightarrow x = 2.$$

Verification: $2ax = 2a(2) \overset{\checkmark}{=} 4a$ ◀

Example 6.10 Solve for x:

$$2ax = 4a^2$$

Solution: ▶

$$2ax = 4a^2 \Rightarrow x = \frac{4a^2}{2a}$$
$$\Rightarrow x = 2a.$$

Verification: $2ax = 2a(2a) \overset{\checkmark}{=} 4a^2.$ ◀

Example 6.11 Solve for x:

$$\frac{x}{3} = -9$$

Solution: ▶

$$\frac{x}{3} = -9 \Rightarrow x = -9(3)$$
$$\Rightarrow x = -27.$$

Verification: $\frac{x}{3} = \frac{-27}{3} \overset{\checkmark}{=} -9.$ ◀

Example 6.12 Solve for x:

$$\frac{x}{a} = a^4$$

Solution: ▶

$$\frac{x}{a} = a^4 \Rightarrow x = a^4(a)$$
$$\Rightarrow x = a^5.$$

Verification: $\frac{x}{a} = \frac{a^5}{a} \overset{\checkmark}{=} a^4.$ ◀

Example 6.13 Solve for x:

$$\frac{ax}{b} = \frac{b}{a}$$

Solution: ▶

$$\frac{ax}{b} = \frac{b}{a} \Rightarrow x = \frac{b}{a} \cdot \frac{b}{a}$$
$$\Rightarrow x = \frac{b^2}{a^2}.$$

Verification: $\frac{ax}{b} = \frac{a}{b} \cdot \frac{b^2}{a^2} \overset{\checkmark}{=} \frac{b}{a}.$ ◀

Example 6.14 Solve for x:

$$\frac{ax}{b} = a^2 b^3$$

Solution: ▶

$$\frac{ax}{b} = a^2 b^3 \Rightarrow x = (a^2 b^3)\frac{b}{a}$$
$$\Rightarrow x = ab^4.$$

Verification: $\frac{ax}{b} = (ab^4)\frac{a}{b} \overset{\checkmark}{=} a^2 b^3$.   ◀

Example 6.15 Solve for x:

$$0 = 12x$$

Solution: ▶

$$0 = 12x \Rightarrow \frac{0}{12} = x$$
$$\Rightarrow x = 0.$$

Verification: $12x = 12(0) \overset{\checkmark}{=} 0$.   ◀

Example 6.16 Solve for x:

$$3x - 5 = 28$$

Solution: ▶

$$3x - 5 = 28 \Rightarrow 3x = 28 + 5$$
$$\Rightarrow x = \frac{33}{3}$$
$$\Rightarrow x = 11.$$

Verification: $3x - 5 = 3(11) - 5 = 33 - 5 \overset{\checkmark}{=} 28$.   ◀

Example 6.17 Solve for x:

$$ax - 2a = 2ab - a$$

Solution: ▶

$$ax - 2a = 2ab - a \Rightarrow ax = 2ab - a + 2a$$
$$\Rightarrow x = \frac{2ab + a}{a}$$
$$\Rightarrow x = 2b + 1.$$

Verification: $ax - 2a = a(2b + 1) - 2a = 2ab + a - 2a \overset{\checkmark}{=} 2ab - a$.   ◀

Supplementary Exercises

Exercise 6.1.1 *Solve for the variable z: $z + 4 = -4$.*

Exercise 6.1.2 *Solve for the variable z: $z - 4 = -4$.*

Exercise 6.1.3 *Solve for the variable z: $4z = -44$.*

Exercise 6.1.4 *Solve for the variable z: $4z = -3$.*

Exercise 6.1.5 *Solve for the variable z:* $\frac{z}{4} = -4$.

Exercise 6.1.6 *Solve for the variable z:* $\frac{2z}{3} = -22$.

Exercise 6.1.7 *Solve for the variable z:* $\frac{2z}{3} + 2 = -22$.

Exercise 6.1.8 *Solve for the variable z:* $\frac{z}{3} = a + 1$.

Exercise 6.1.9 *Solve for the variable z:* $5z - 3 = 22$.

Exercise 6.1.10 *Solve for the variable z:* $\frac{bz}{a+1} - b^3 = ab$.

Exercise 6.1.11 *Solve for the variable N:* $3N = 3A$.

Exercise 6.1.12 *Solve for the variable N:* $\frac{N}{A} = A^2 + 1$.

Exercise 6.1.13 *Solve for the variable N:* $\frac{XN}{Y} = \frac{Y}{X}$.

Exercise 6.1.14 *Solve for the variable N:* $N - 5A = 7A + A^2$.

Exercise 6.1.15 *Solve for the variable N:* $N + 8X = -9X$.

Exercise 6.1.16 *Solve for the variable N:* $\frac{N}{A+2} = A + 1$.

Exercise 6.1.17 *Solve for the variable N:* $\frac{A^2 N}{B^3} = A^2 B^4$.

Exercise 6.1.18 *Solve for the variable N:* $\frac{7N}{A} = \frac{14}{A^2}$.

Exercise 6.1.19 *Solve for the variable N:* $N + 2 = A + 14$.

6.2 MISCELLANEOUS LINEAR EQUATIONS

Example 6.18 Solve for x:

$$\frac{x}{2} - \frac{x}{3} = \frac{x}{4} + \frac{x}{2}$$

Solution: ▶ *The denominators are 3, 4 and 2, and their least common multiple is (3, 4, 2) = 12. We multiply both sides of the equation by this least common multiple, and so*

$$\frac{x}{2} - \frac{x}{3} = \frac{x}{4} + \frac{x}{2} \Rightarrow 12\left(\frac{x}{2} - \frac{x}{3}\right) = 12\left(\frac{x}{4} + \frac{1}{2}\right)$$
$$\Rightarrow 6x - 4x = 3x + 6$$
$$\Rightarrow 2x = 3x + 6$$
$$\Rightarrow 2x - 3x = 6$$
$$\Rightarrow -x = 6$$
$$\Rightarrow x = -6$$
◀

Example 6.19 Solve for a:

$$\frac{b-a}{6} = \frac{b-a}{15} - b.$$

Solution: ▶ *The denominators are* **6** *and* **15**, *and their least common multiple is* **(6, 15) = 30**. *We multiply both sides of the equation by this least common multiple, and so*

$$\frac{b-a}{6} = \frac{b-a}{15} - b \Rightarrow 30\left(\frac{b-a}{6}\right) = 30\left(\frac{b-a}{15} - b\right)$$
$$\Rightarrow 5(b-a) = 2(b-a) - 30b$$
$$\Rightarrow 5b - 5a = 2b - 2a - 30b$$
$$\Rightarrow -5a + 2a = 2b - 5b - 30b$$
$$\Rightarrow -3a = -33b$$
$$\Rightarrow a = 11b. \qquad\blacktriangleleft$$

Example 6.20 Solve for x:

$$\frac{2x-1}{3} = \frac{1-3x}{2}.$$

Solution: ▶ *Cross–multiplying,*

$$\frac{2x-1}{3} = \frac{1-3x}{2} \Rightarrow 2(2x-1) = 3(1-3x)$$
$$\Rightarrow 4x - 2 = 3 - 9x$$
$$\Rightarrow 4x + 9x = 3 + 2$$
$$\Rightarrow 13x = 5$$
$$\Rightarrow x = \frac{5}{13}. \qquad\blacktriangleleft$$

Example 6.21 Write the infinitely repeating decimal $0.3\overline{45} = 0.345454545\ldots$ as the quotient of two natural numbers.

Solution: ▶ *The trick is to obtain multiples of* $x = 0.345454545\ldots$ *so that they have the same infinite tail, and then subtract these tails, cancelling them out.*[1] *Observe that*

$10x = 3.45454545\ldots; 1000x = 345.454545\ldots \Rightarrow 1000x - 10x = 342 \Rightarrow x = \frac{342}{990} = \frac{19}{55}. \qquad\blacktriangleleft$

Example 6.22 Find the fraction that represents the repeating decimal

$$4.3\overline{21} = 4.32121212121\ldots$$

Solution: ▶ *Put* $x = 4.32121212121\ldots$ *Then* $10x = 43.21212121\ldots$ *and* $1000x = 4321.2121212121\ldots$ *Therefore*

$$1000x - 10x = 4321.2121212121\ldots - 43.212121212121 \ldots 4278.$$

Finally

$$x = \frac{4278}{990} = \frac{2139}{495} = \frac{713}{165}.$$

---

[1] That this cancellation is meaningful depends on the concept of *convergence*.

Example 6.23 Find the sum of the first hundred positive integers, that is, find

$$1 + 2 + 3 + \cdots + 99 + 100.$$

Solution: ▶ *Notice that this is Example 1.5. Put*

$$x = 1 + 2 + 3 + \cdots + 99 + 100.$$

The crucial observation is that adding the sum forwards is the same as adding the sum backwards, therefore

$$x = 100 + 99 + \cdots + 3 + 2 + 1.$$

Adding,

$$
\begin{array}{rcccccccc}
x & = & 1 & + & 2 & + \ldots + & 99 & + & 100 \\
x & = & 100 & + & 99 & + \ldots + & 2 & + & 1 \\
\hline
2x & = & 101 & + & 101 & + \ldots + & 101 & + & 101 \\
& = & 100 \cdot 101,
\end{array}
$$

therefore,

$$x = \frac{100 \cdot 101}{2} = 50 \cdot 101 = 5050.$$ ◀

Example 6.24 Consider the arithmetic progression

$$2, 7, 12, 17, \ldots.$$

Is **302** in this progression? Is **303** in this progression?

Solution: ▶ *Observe that we start with* **2** *and keep adding* **5**. *Thus*

$$2 = 2 + 5 \cdot 0, 7 = 2 + 5 \cdot 1, 12 = 2 + 5 \cdot 2, 17 = 2 + 5 \cdot 3, \ldots$$

so the general term has the form **5n + 2**, *which are the numbers leaving remainder* **2** *when divided by* **5**. *Since*

$$5n + 2 = 302 \Rightarrow 5n = 300 \Rightarrow n = 60,$$

is an integer, **302** *is in this progression. Since*

$$5n + 2 = 303 \Rightarrow 5n = 301 \Rightarrow n = \frac{301}{5},$$

is an not integer, **303** *is not in this progression.* ◀

Some equations are nonlinear, but may be reduced to a linear equation. Sometimes it is necessary to simplify the expressions involved before solving the equation, as seen in the following examples:

Example 6.25 Solve for x:

$$(3x + 1)^2 + 6 + 18(x + 1)^2 = 9x(3x - 2) + 65.$$

Solution: ▶

$$(3x+1)^2 + 6 + 18(x+1)^2 = 9x(3x-2) + 65$$
$$\Rightarrow 9x^2 + 6x + 1 + 6 + 18x^2 + 36x + 18 = 27x^2 - 18x + 65$$
$$\Rightarrow 27x^2 + 42x + 25 = 27x^2 - 18x + 65$$
$$\Rightarrow 27x^2 + 42x - 27x^2 + 18x = 65 - 25$$
$$\Rightarrow 60x = 40$$
$$\Rightarrow x = \frac{2}{3}.$$

◀

Example 6.26 Solve for x:

$$(2x+1)(2x+6) - 7(x-2) = 4(x+1)(x-1) - 9x$$

Solution: ▶

$$(2x+1)(2x+6) - 7(x-2) = 4(x+1)(x-1) - 9x$$
$$\Rightarrow 4x^2 + 14x + 6 - 7x + 14 = 4x^2 - 4 - 9x$$
$$\Rightarrow 4x^2 + 7x + 20 = 4x^2 - 9x - 4$$
$$\Rightarrow 4x^2 + 7x - 4x^2 + 9x = -4 - 20$$
$$\Rightarrow 16x = -24$$
$$\Rightarrow x = -\frac{3}{2}.$$

◀

Example 6.27 Solve for x:

$$\frac{2}{x+1} = \frac{3}{x}.$$

Solution: ▶ *Cross-multiplying,*

$$\frac{2}{x+1} = \frac{3}{x} \Rightarrow 2x = 3(x+1) \Rightarrow 2x = 3x + 3 \Rightarrow 2x - 3x = 3 \Rightarrow -x = 3 \Rightarrow x = -3.$$

◀

Supplementary Exercises

Exercise 6.2.1 *Solve for x:* $2(3x-4) - 4(2-3x) = 1.$

Exercise 6.2.2 *Solve for x:* $x - \frac{x}{2} - \frac{x}{3} = 1.$

Exercise 6.2.3 *Solve for x:* $\frac{x-2}{2} = \frac{3-x}{3}.$

Exercise 6.2.4 *Solve for x:* $\frac{x}{a} - 1 = 2.$

Exercise 6.2.5 *Solve for x:* $\frac{ax}{b} = a.$

Exercise 6.2.6 *Solve for x:* $ax + b = c.$

Exercise 6.2.7 *Solve for x:* $\frac{x+a}{2} = 2x + 1.$

Exercise 6.2.8 *Solve for* x: $= \dfrac{x+1}{2} - \dfrac{x+2}{3} = \dfrac{x-1}{4}$.

Exercise 6.2.9 *Solve for* x: $\dfrac{a}{x} = b$.

Exercise 6.2.10 *Solve for* x: $\dfrac{ab}{cx} = d$.

Exercise 6.2.11 *Solve for* x: $\dfrac{3}{x-2} = 1$.

Exercise 6.2.12 *Solve for* x: $\dfrac{3}{x-2} = \dfrac{2}{x+3}$.

Exercise 6.2.13 *Solve for* x: $2(3-x) = 3x - 4$

Exercise 6.2.14 *Solve for* x: $(2-x)(x+3) = -x(x-4)$

Exercise 6.2.15 *Solve for* x: $(x-a)b = (b-x)a$.

Exercise 6.2.16 *Solve for* x: $\dfrac{x-3}{3} - \dfrac{x-2}{2} = 6$.

Exercise 6.2.17 *Write the infinitely repeating decimal*

$$0.\overline{123} = 0.123123123... \text{ as the quotient of two positive integers.}$$

6.3 WORD PROBLEMS

Example 6.28 Find two numbers whose sum is **28**, and whose difference is **4**.

Solution: ▶ *Let* x *be one of the numbers, then the other number is* $28 - x$. *Then we have*

$$x - (28 - x) = 4 \Rightarrow 2x - 28 = 4$$

$$\Rightarrow x = 16.$$

The numbers are $x = 16$ *and* $28 - x = 28 - 16 = 12$. ◀

Example 6.29 Divide **$47** between Peter, Paul, and Mary, so that Peter may have **$10** more than Paul, and Paul **$8** more than Mary.

Solution: ▶ *Let* p *be Paul's amount in dollars. Then Peter has* $p + 10$ *dollars and Mary has* $p - 8$ *dollars. Then we have*

$$p + (p + 10) + (p - 8) = 47 \Rightarrow 3p + 2 = 47$$

$$\Rightarrow 3p = 45$$

$$\Rightarrow p = 15.$$

Thus Paul has **$15**, *Peter has* **$25**, *and Mary has* **$7**. ◀

Example 6.30 The sum of three consecutive odd integers is **609**. Find the numbers.

Solution: ▶ *Let the numbers be* $x - 2, x, x + 2$. *Then*

$$(x - 2) + x + (x + 2) = 609 \Rightarrow 3x = 609$$

$$\Rightarrow x = 203$$

The numbers are $x - 2 = 201, x = 203,$ *and* $x + 2 = 205$. ◀

Example 6.31 A glass of milk costs **40** cents more than a loaf of bread, but **50** cents less than a glass of juice. If the cost of the three items, in cents, is **730**, what is the price of each item, in cents?

> **Solution:** ▶ *Let **m** be the price of a glass of milk in cents. Then bread costs **m − 40** cents and juice costs **m + 50** cents. This gives*
>
> $$m+(m-40)+(m+50)=730 \Rightarrow 3m+10=730 \Rightarrow m=240.$$
>
> *Thus milk costs **240** cents, bread costs **200** cents and juice costs **290** cents.* ◀

Example 6.32 Currently, the age of a father is four times the age of his son, but in **24** years from now it will only be double. Find their ages.

> **Solution:** ▶ *Let **s** be the current age of the son. Then the current age of the father is **4s**. In* **24** *years the son will be **s + 24** and the father will be **4s + 24** and we will have*
>
> $$4s+24=2(s+24).$$
>
> *This gives*
>
> $$4s+24=2s+48 \Rightarrow 4s-2s=48-24 \Rightarrow s=12.$$
>
> *Thus the son is currently **12** years old and the father is currently **48** years old.* ◀

Supplementary Exercises

Exercise 6.3.1 *Six times a number increased by **11** is equal to **65**. Find the number.*

Exercise 6.3.2 *Find a number which when multiplied by **11** and then diminished by **18** is equal to **15**.*

Exercise 6.3.3 *If **3** is added to a number, and the sum multiplied by **12**, the result is **84**. Find the number.*

Exercise 6.3.4 *The sum of eleven consecutive integers is **2002**. Find them.*

Exercise 6.3.5 *One number exceeds another by **3**, and their sum is **27**. Find them.*

Exercise 6.3.6 *Find two numbers whose sum is **19**, such that one shall exceed twice the other by **1**.*

Exercise 6.3.7 *Split $380 among Peter, Paul and Mary, so that Paul has $30 more than Peter, and Mary has $20 more than Paul.*

Exercise 6.3.8 *How much pure water should be added to **100** grams of **60%** acid solution to make a **20%** acid solution?*

Exercise 6.3.9 *Jane's age is twice Bob's age increased by **3**. Bill's age is Bob's age decreased by **4**. If the sum of their ages is **27**, how old is Bill?*

Exercise 6.3.10 *The average of six numbers is **4**. A seventh number is added and the new average increases to **5**. What was the seventh number?*

Exercise 6.3.11 *Bill currently has five times as much money as Bob. If he gives $20 to Bob, then Bill will only have four times as much. Find their current amounts.*

Exercise 6.3.12 *Find a number so that six sevenths of it exceed four fifths of it by **2**.*

Exercise 6.3.13 *The difference between two numbers is **8**. If we add **2** to the larger number we obtain **3** times the smaller one. Find the numbers.*

Exercise 6.3.14 *Find two numbers whose difference is **10**, and whose sum equals twice their difference.*

Exercise 6.3.15 *I bought a certain amount of avocados at four for $2; I kept a fifth of them, and then sold the rest at three for $2. If I made a profit of $2, how many avocados did I originally buy?*

Exercise 6.3.16 *Find a number whose fourth, sixth, and eighth of its value add up to 13.*

Exercise 6.3.17 *A fifth of the larger of two consecutive integers exceeds a seventh of the smaller by 3. Find the integers.*

Exercise 6.3.18 *I bought a certain number of oranges at three for a dollar and five sixths of that number at four for a dollar. If I sold all my oranges at sixteen for six dollars, I would make a profit of three and a half dollars. How many oranges did I buy?*

Exercise 6.3.19 *A piece of equipment is bought by a factory. During its first year, the equipment depreciates a fifth of its original price. During its second year, it depreciates a sixth of its new value. If its current value is $56,000, what was its original price?*

Exercise 6.3.20 *The difference of the squares of two consecutive integers is 665. Find the integers.*

Exercise 6.3.21 *In how many different ways can one change 50 cents using nickels, dimes or quarters?*

Exercise 6.3.22 *Five burglars stole a purse with gold coins. The burglars each took different amounts according to their meanness, with the meanest among the five taking the largest amount of coins and the meekest of the five taking the least amount of coins. Unfair sharing caused a fight that was brought to an end by an arbitrator. He ordered that the meanest burglar should double the shares of the other four burglars below him. Once this was accomplished, the second meanest burglar should double the shares of the three burglars below him. Once this was accomplished, the third meanest burglar should double the shares of the two burglars below him. Once this was accomplished, the fourth meanest burglar should double the shares of the meekest burglar. After this procedure was terminated, each burglar received the same amount of money. How many coins were in a purse if the meanest of the burglars took 240 coins initially?*

QUADRATIC EQUATIONS IN ONE-VARIABLE EXPRESSIONS

This chapter applies the mathematical operations of addition, subtraction, multiplication, and division in algebraic quadratic equations. We introduce this concept through these sections: *Quadratic Equations, Complete the Square,* and *Quadratic Formula.*

7.1 QUADRATIC EQUATIONS

Definition 7.1 A *quadratic equation* in the unknown x is an equation in which a variable is squared or written to the second power of the form

$$ax^2 + bx + c = 0$$

where a $\neq$ 0, b, c are real numbers.

We will circumscribe ourselves to the study of quadratic trinomials $ax^2 + bx + c$ that are amenable to the factorizations studied earlier. When a variable is squared in an equation, and if the product of two real numbers is zero, then at least one of them must be zero. In symbols,

$$ab = 0 \Rightarrow a = 0 \quad \text{or} \quad b = 0.$$

Quadratic equations are solved by factoring their terms. Two expressions when multiplied will equal the quadratic equation.

Example 7.1 Solve the quadratic equation $x^2 - x = 0$ for x.

Solution: ▶ *We first factor the expression $x^2 - x = x(x - 1)$. Now, $x(x - 1) = 0$ is the product of two real numbers giving 0, therefore either $x = 0$ or $x - 1 = 0$. Thus either $x = 0$ or $x = 1$. One can easily verify now that $0^2 - 0 = 0$ and $1^2 - 1 = 0$.* ◀

Example 7.2 Solve the quadratic equation $x^2 + 2x - 3 = 0$ for x.

Solution: ▶ *Factoring,*

$$x^2 + 2x - 3 = 0 \Rightarrow (x + 3)(x - 1) = 0$$

When a quadratic equation is set to equal zero and the factors are known, you can find the solutions by appling the zero-product rule, where any number when multiplied by 0 is equal to 0.

$$\Rightarrow x + 3 = 0 \quad \text{or} \quad x - 1 = 0$$

$$\Rightarrow x = -3 \quad \text{or} \quad x = 1$$
◀

Check: *Substitute each solution (x = −3 and x = 4) into the original equation. Both solutions equal zero (−3 + 3 = 0, and 1 − 1 = 0).*

To multiply expressions, multiply each term in the second expression by each term in the first expression. The acronym to remember is <u>FOIL</u>, where you multiply the terms in the following order: <u>F</u>irst, <u>O</u>utside, <u>I</u>nside, <u>L</u>ast.

$$(x+3)(x-1) = x(x) + x(-1) + 3(x) + 3(-1) = x^2 - x + 3x - 3 = x^2 + 2x - 3$$

Example 7.3 Solve the quadratic equation $8x^2 - 2x = 15$ for x.

Solution: ▶ *First, transform the equation into $8x^2 - 2x - 15 = 0$. Now, factoring,*

$$8x^2 - 2x - 15 \Rightarrow (2x - 3)(4x + 5) = 0$$

$$\Rightarrow 2x - 3 = 0 \quad \text{or} \quad 4x + 5 = 0$$

$$\Rightarrow x = \frac{3}{2} \quad \text{or} \quad x = -\frac{5}{4}. \qquad \blacktriangleleft$$

Some equations can be reduced to quadratic equations after proper massaging

Example 7.4 Solve for x: $\frac{x-4}{3} = \frac{4}{x}$.

Solution: ▶ *Cross-multiplying,*

$$\frac{x-4}{3} = \frac{4}{x} \Rightarrow x(x-4) = 12$$

$$\Rightarrow x^2 - 4x - 12 = 0$$

$$\Rightarrow (x - 6)(x + 2) = 0$$

$$\Rightarrow x = 6 \text{ or } x = -2. \qquad \blacktriangleleft$$

The argument is that a product of real numbers is zero if and only if at least one of its factors can be used with equations of higher degree.

Example 7.5 Solve for x: $x^3 - 4x = 0$.

Solution: ▶ *Factoring a common factor and then a difference of squares,*

$$x^3 - 4x = 0 \Rightarrow x(x^2 - 4) = 0$$

$$\Rightarrow x(x - 2)(x + 2) = 0$$

$$\Rightarrow x = 0 \quad \text{or} \quad x = +2 \quad \text{or} \quad x = -2. \qquad \blacktriangleleft$$

Supplementary Exercises

Exercise 7.1.1 *Solve for x: $x^2 - 4 = 0$.*

Exercise 7.1.2 *Solve for x: $x^2 - x - 6 = 0$.*

Exercise 7.1.3 *Solve for x: $x^2 + x - 6 = 0$.*

Exercise 7.1.4 *Solve for x: $x^2 - 4x = 5$.*

Exercise 7.1.5 *Solve for x: $x^2 = 1$.*

Exercise 7.1.6 *If eggs had cost x cents less per dozen, it would have cost 3 cents less for $x + 3$ eggs than if they had cost x cents more per dozen. What is x?*

Exercise 7.1.7 *A group of people rents a bus for an excursion for $2,300, with each person paying an equal share. The day of the excursion, it is found that six participants do not show up, which increases the share of those present by $7.50. What was the original number of people renting the bus?*

Exercise 7.1.8 *Let*

$$x = 6 + \cfrac{1}{6 + \cfrac{1}{6 + \cfrac{1}{6 + \cfrac{1}{6 + \cfrac{1}{\ddots}}}}},$$

where there are an infinite number of fractions. Prove that $x^2 - 6x - 1 = 0$.

Exercise 7.1.9 *Assume that there is a positive real number* x *such that*

$$x^{x^{x^{x^{\cdot^{\cdot^{\cdot}}}}}} = 2,$$

where there is an infinite number of x's*. What is the value of* x?

Exercise 7.1.10 *Factor:*

$$x^2 + 20x + 19,$$

Exercise 7.1.11 *Factor:*

$$x^2 + 14x + 13,$$

Exercise 7.1.12 *Factor:*

$$z^2 + 14z + 49,$$

Exercise 7.1.13 *Factor:*

$$x^2 + 2x + 1,$$

Exercise 7.1.14 *Factor:*

$$x^2 + 6x + 9,$$

Exercise 7.1.15 *Factor:*

$$x^2 - 10x + 25,$$

Exercise 7.1.16 *Factor:*

$$x^2 - 6x + 9,$$

Exercise 7.1.17 *Factor:*

$$x^2 - 12x + 36,$$

Exercise 7.1.18 *Solve for* z. *Write your answers as whole numbers and in the simplest form.*

$$z(z - 2) = 0,$$

Exercise 7.1.19 *Solve for* ***n***. *Write your answers in the simplest form.*

$$(3n+4)(n-2)=0,$$

Exercise 7.1.20 *Solve for* ***t***. *Write your answers in the simplest form.*

$$(5t-9)(t+2)=0,$$

Exercise 7.1.21 *Solve for* ***x***. *Write your answers in the simplest form.*

$$(5x+9)(x-2)=0,$$

Exercise 7.1.22 *Solve for* ***n***. *Write your answers in the simplest form.*

$$(2n+3)(n-13)=0,$$

7.2 COMPLETE THE SQUARE

With quadratic equations like x^2+bx, you can complete the square by adding $\left(\frac{b}{2}\right)^2$.

Example 7.6 *Complete the square by filling in the number that makes the polynomial a perfect square.*

$$x^2+44x+\underline{\hspace{1cm}}$$

Solution: ▶ *Add* $\left(\frac{b}{2}\right)^2$ *to complete the square.*

$$x^2+44x+\left(\frac{b}{2}\right)^2$$

$$x^2+44x+\left(\frac{44}{2}\right)^2 \quad \textbf{substitute in } b=44$$

$$x^2+44x+22^2$$

$$x^2+44x+484. \qquad\qquad\blacktriangleleft$$

This quadratic equation can be written in a complete square, $(x+22)^2$. The perfect square was complete with the number **484**.

Supplementary Exercises

Exercise 7.2.1 *Complete the square by filling in the number that makes the polynomial a perfect square.*

$$x^2+30x+\underline{\hspace{1cm}}$$

Exercise 7.2.2 *Complete the square by filling in the number that makes the polynomial a perfect square.*

$$x^2-6x+\underline{\hspace{1cm}}$$

Exercise 7.2.3 *Complete the square by filling in the number that makes the polynomial a perfect square.*

$$h^2+10h+\underline{\hspace{1cm}}$$

Exercise 7.2.4 *Complete the square by filling in the number that makes the polynomial a perfect square.*

$$x^2 - 16x + \underline{\qquad}$$

Exercise 7.2.5 *Complete the square by filling in the number that makes the polynomial a perfect square.*

$$p^2 + 36p + \underline{\qquad}$$

Exercise 7.2.6 *Complete the square by filling in the number that makes the polynomial a perfect square.*

$$x^2 + 28x + \underline{\qquad}$$

Exercise 7.2.7 *Complete the square by filling in the number that makes the polynomial a perfect square.*

$$x^2 - 26x + \underline{\qquad}$$

Exercise 7.2.8 *Complete the square by filling in the number that makes the polynomial a perfect square.*

$$h^2 - 10h + \underline{\qquad}$$

Exercise 7.2.9 *Complete the square by filling in the number that makes the polynomial a perfect square.*

$$x^2 + 2x + \underline{\qquad}$$

Exercise 7.2.10 *Complete the square by filling in the number that makes the polynomial a perfect square.*

$$x^2 - 2x + \underline{\qquad}$$

7.3 QUADRATIC FORMULA

A quadratic equation $ax^2 + bx + c = 0$, where $a \neq 0$, can be solved using the following quadratic formula:

$$x = \frac{-b \pm \sqrt{b^2 - 4ac}}{2a}.$$

Example 7.7 *Use the quadratic formula to solve* $4x^2 - 9x + 3 = 0$,

$$x = \frac{-b \pm \sqrt{b^2 - 4ac}}{2a} = \frac{-(-9) \pm \sqrt{(-9)^2 - 4(4)(3)}}{2(4)}$$

$$\frac{9 \pm \sqrt{81 - 48}}{8} \rightarrow \frac{9 \pm \sqrt{33}}{8}$$

$$x = \frac{9 + \sqrt{33}}{8} \quad \text{or} \quad x = \frac{9 - \sqrt{33}}{8}. \text{ Split into } + \text{ or } -$$

$$x \approx 1.84 \quad \text{or} \quad x \approx 0.41.$$

Simplify and round to nearest hundredth

Example 7.8 *Use the quadratic formula to solve* $p^2 - 8p + 6 = 0$,

$$x = \frac{b \pm \sqrt{b^2 - 4ac}}{2a} = \frac{8 \pm \sqrt{(8)^2 - 4(1)(6)}}{2(1)}$$

$$\frac{8 \pm \sqrt{64 - 24}}{2} \rightarrow \frac{8 \pm \sqrt{40}}{2} \rightarrow \frac{8 \pm 2\sqrt{10}}{2} \rightarrow 4 \pm \sqrt{10}$$

$$x = 4 + \sqrt{10} \quad \text{or} \quad x = 4 - \sqrt{10} \text{ Split into } + \text{ or } -$$

$$x \approx 7.16 \quad \text{or} \quad x \approx 0.84.$$

Simplify and round to nearest hundredth

Example 7.9 *Use the quadratic formula to solve* $6n^2 - 2n - 8 = 0$,

$$x = \frac{-b \pm \sqrt{b^2 - 4ac}}{2a} = \frac{-(-2) \pm \sqrt{(-2)^2 - 4(6)(-8)}}{2(6)}$$

$$\frac{2 \pm \sqrt{4 + 192}}{12} \rightarrow \frac{2 \pm \sqrt{196}}{12} \rightarrow \frac{2 \pm 14}{12}$$

$$x = \frac{16}{12} \quad \text{or} \quad x = -\frac{12}{12} \text{ Split into } + \text{ or } -$$

$$x = \frac{4}{3} \quad \text{or} \quad x = -1.$$

Supplementary Exercises

Exercise 7.3.1 *Find the solution for the following quadratic equation using the quadratic formula.*

$$p^2 + 7p - 9 = 0,$$

Exercise 7.3.2 *Find the solution for the following quadratic equation using the quadratic formula.*

$$4x^2 - 3x - 8 = 0,$$

Exercise 7.3.3 *Find the solution for the following quadratic equation using the quadratic formula.*

$$3f^2 - 7f - 7 = 0,$$

Exercise 7.3.4 *Find the solution for the following quadratic equation using the quadratic formula.*

$$2z + 7z + 5 = 0,$$

Exercise 7.3.5 *Find the solution for the following quadratic equation using the quadratic formula.*

$$2q^2 + 4q + 2 = 0,$$

Exercise 7.3.6 *Find the solution for the following quadratic equation using the quadratic formula.*

$$2n^2 - 9n - 4 = 0,$$

Exercise 7.3.7 *Find the solution for the following quadratic equation using the quadratic formula.*

$$4x^2 - 3x - 9 = 0,$$

Exercise 7.3.8 *Find the solution for the following quadratic equation using the quadratic formula.*

$$5u^2 + 9u - 7 = 0,$$

Exercise 7.3.9 *Find the solution for the following quadratic equation using the quadratic formula.*

$$2w^3 - 9w^2 + 8w = 0,$$

Exercise 7.3.10 *Find the solution for the following quadratic equation using the quadratic formula.*

$$2m^3 + 8m^2 + 2m = 0,$$

PART 5

INEQUALITIES

CHAPTER 8

LINEAR INEQUALITIES

This chapter applies the mathematical operations of addition, subtraction, multiplication, and division in algebraic linear inequality equations. We introduce linear inequalities in sections covering *Intervals, One-Variable Linear Inequalities*, and *Using One-Variable Linear Inequalities to Solve Word Problems*.

8.1 INTERVALS

Linear inequalities are statements that mathematically express that two expressions or numbers are not equal in value. Solving linear inequalities is very similar to solving linear equations discussed earlier. The symbols below are used to show the relationship between the two sides of a linear inequality statement.

$<$	(is less than)	$y < 3$
$>$	(is greater than)	$x > 10$
$\leq$	(is less than or equal to)	$7 \leq a$
$\geq$	(is greater than or equal to)	$z \geq -5$

We will simply state the useful properties for the symbol $<$, as the same properties hold for the other three symbols ($>$, $\leq$, and $\geq$).

Axiom 8.1 Let a, b, c be real numbers. Then

1. If $a < b$ and $b < c$, then $a < c$, that is, inequalities are *transitive*.

2. If $a < b$ and $c < d$, then $a + c < b + d$, that is, addition preserves the inequality sense.

3. If $a < b$ and $c > 0$, then $ac < bc$, that is, multiplication by a positive factor preserves the inequality sense.

4. If $a < b$ and $c < 0$, then $ac > bc$, that is, multiplication by a negative factor reverses the inequality sense.

Definition 8.1 An interval is a set I of real numbers with the following property: for any two different elements x, y in I, any number a in between x and y also belongs to I.

☞ *From the definition we see intervals have infinitely many elements. For example, consider the interval in Figure 8.11. Every number between -2 and 1 is in this interval, so in particular, -1.5 belongs to it, so does -1.51, -1.501, -1.5001, etc.*

Some examples of intervals are shown in Figures 8.1 through 8.12. If we define the *length* of the interval to be the difference between its final point and its initial point, then intervals can be finite, like those in Figures 8.9 through 8.12 or infinite, like those in Figures 8.1 through 8.8. In either case, intervals contain an infinite number of elements.

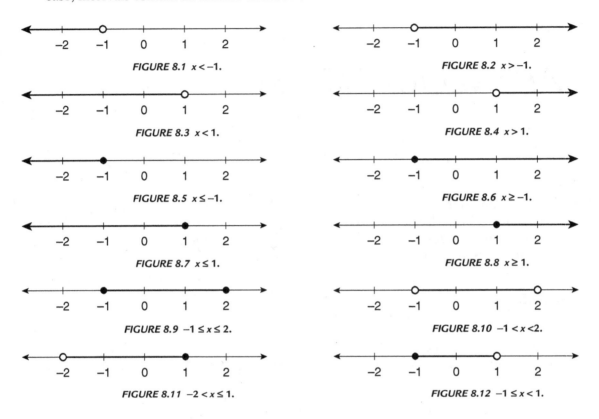

FIGURE 8.1 $x < -1$.

FIGURE 8.2 $x > -1$.

FIGURE 8.3 $x < 1$.

FIGURE 8.4 $x > 1$.

FIGURE 8.5 $x \leq -1$.

FIGURE 8.6 $x \geq -1$.

FIGURE 8.7 $x \leq 1$.

FIGURE 8.8 $x \geq 1$.

FIGURE 8.9 $-1 \leq x \leq 2$.

FIGURE 8.10 $-1 < x < 2$.

FIGURE 8.11 $-2 < x \leq 1$.

FIGURE 8.12 $-1 \leq x < 1$.

8.2 ONE-VARIABLE LINEAR INEQUALITIES

We are now interested in solving linear inequalities in one-variable. The process will closely resemble that for solving linear equations in one-variable, with the caveat that if we ever divide or multiply by a negative quantity, then the sense of the inequality changes. This can be done by simply isolating the unknown variable on the left side of the inequality and moving the rest of the expresson to the right side of the inequality. To do this, make sure you perform the same mathematical operation on both sides of the inequality

Example 8.1 Solve the inequality and graph its solution set.

$$2x - 3 < -13$$

Solution: ▶ *Add* **3** *to both sides of the inequality*

$$2x - 3 + 3 < -13 + 3$$

$$2x < -10$$

▶ *Divide by* **2**, *which is a positive quantity. The sense of the inequality is retained, and we gather,*

$$\frac{2x}{2} < \frac{-10}{2}$$

$$x < -5$$

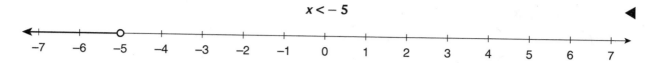

FIGURE 8.13 Graphical solution for example 267: $x < -5$.

Example 8.2 Solve the inequality and graph its solution set.

$$-2x - 3 \leq -13$$

Solution: ▶ *Add* **3** *to both sides of the inequality*

$$-2x - 3 + 3 \leq -13 + 3$$

$$-2x \leq -10$$

▶ *Divide by* **2**, *which is a positive quantity. The sense of the inequality is reversed, and we gather,*

$$\frac{-2x}{-2} \geq \frac{-10}{-2}$$
$$x \geq -5$$

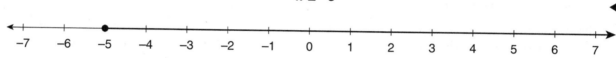

FIGURE 8.14 Graphical solution for example 8.2 $x \geq -5$.

Supplementary Exercises

Exercise 8.2.1 *Find the set of all real numbers* **x** *satisfying the inequality and graph its solution set.*

$$x - \frac{x - 6}{4} < 2 + x,$$

Exercise 8.2.2 *Find the set of all real numbers* **x** *satisfying the inequality and graph its solution set.*

$$4(x + 1) \leq x - 5,$$

Exercise 8.2.3 *Find the set of all real numbers* **x** *satisfying the inequality and graph its solution set.*

$$\frac{x - 1}{3} - \frac{1 - x}{2} > 0,$$

8.3 USING ONE-VARIABLE LINEAR INEQUALITIES TO SOLVE WORD PROBLEMS

Inequalities can be used to solve real-life problems. The procedure to solve any word problems that have a range of answers is to set the problems up with inequalities, as follows:

1. Assign a variable to the unknown numbers in the word problem.

2. Set a relationship between the variable and the other numbers in the word problem.

3. Use the appropriate inequality expression to define the relationship between the unknown variable and the other numbers.

Example 8.3 Mark wants to save at least **$1,200** for a down payment on a used vehicle. What is the smallest amount of money he can set aside each month so that he can reach his goal in **6** months?

Solution: ▶ *Let* **x** *be the amount of money that Mark will save each month, then* **6x** *represents the amount he will save in* **6** *months. His savings must be greater than or equal to* **$1,200**.

$$6x \geq 1200$$

$$\frac{6x}{6} \geq \frac{1200}{6}$$

$$x \geq 200$$

Solution: ▶ *Mark should set a goal of saving a minimum of $200 per month.*

Check: ▶ *Replace x with* **200** *or* **6(200) = 1200**. *If Mark saves* **$200** *for* **6** *months, he will have* **$1,200**, *which is enough for the down payment on a used vehicle.*

It is not necessary to use linear inequality statement to solve this type of problem. Most people will solve the problem using equations or their common sense. You know that **$200** *times* **6** *equals* **$1,200**. *Using common sense, you know that Mark can save even more if he puts away even a larger amount in savings each month. Thus, his savings must be greater than or equal to* **$200**.

There are times when you need to set up word problems in the best way possible using linear inequality statements. The best setup for the following example is to use a distance formula.

Example 8.4 Andreya computes to work from Raleigh to Chapel Hill daily, a distance of **30** miles. If she drives no faster than **60** miles per hour, what is the total time in hours that the trip will take her each day?

Solution: ▶ *To solve this problem, use the distance formula* **d = rt**, *where* **d** *is the distance,* **r** *is the rate of travel (speed), and* **t** *is the time in hours. The distance Andreya travels is* **30** *miles* (**d = 30**) *and the rate of travel is* **60** *miles per hours* (**r = 60**). *Therefore* **60t** *represents distance. Her traveling speed must be less than or equal to* **60**.

$$60t \geq 30$$

$$\frac{60t}{60} \geq \frac{30}{60}$$

$$t \geq \frac{30}{60}$$

$$t \geq \frac{1}{2}$$

In other words, if Andreya drives any slower than **60** *miles per hour, her commute will take longer than* $\frac{1}{2}$ *hour or* **30** *minute,* **60t ≥ 30**. ◀

Supplementary Exercises

Exercise 8.3.1 *An unknown number is doubled and decreased in its value by 50 is less than 100. Which of the following is the correct statement?*

(A) **x < 75**   (C) **x <25**

(B) **x >75**   (D) *x >25*

Exercise 8.3.2 *Dalya works at least* **40** *hours a week as a cashier at a department store, making* **$10.75** *per hour. Write the inequality statement that expresses her biweekly take home pay.*

Exercise 8.3.3 *The perimeter of a square is larger than* **72** *inches. Set up the inequality expression to solve for the size of the square (**s**) in inches?*

Exercise 8.3.4 *A tennis ball is projected upward in the air with an intial velocity of* **V = 75 − 30t**, *where* **t** *is the time it travels in seconds. When will the velocity of the ball reach between* **30** *and* **60** *feet per second?*

Exercise 8.3.5 *Robert invests $10,000 into a financial portfolio, where 3% of his investment is placed into a 3% simple interest account. The remainder of his investment is placed in mutual funds that return an average of 11% interest. What is the least amount of investment that Robert can invest in mutual funds to receive $2,700 interest in a year?*

8.4 COMPOUND LINEAR INEQUALITIES

Compound linear inequalities are simply inequality expressions that have more than one set of constraints. The process will closely resemble that for solving linear inequality equations in one-variable, with the caveat that if we divide or multiply by a negative quantity, then the sense of the inequality changes. This can be done by simply isolating the unknown variable on the left side of the inequality and moving the rest of the expresson to the right side of the inequality. To do this, make sure you perform the same mathematical operation on both sides of the inequality.

Example 8.5 Solve the compound inequalities and graph their solution set.

$$x - 15 < -1 \quad \text{or} \quad x - 13 \geq 6$$

Solution: ▶ *Treat each inequality expression separately. For example, add* **15** *to both sides of the inequality for the first expression and* **13** *to the other expression.*

$$x - 15 + 15 < -1 + 15 \quad \text{or} \quad x - 13 + 13 \geq 6 + 13$$

$$x < 14 \quad \text{or} \quad x \geq 19. \qquad \blacktriangleleft$$

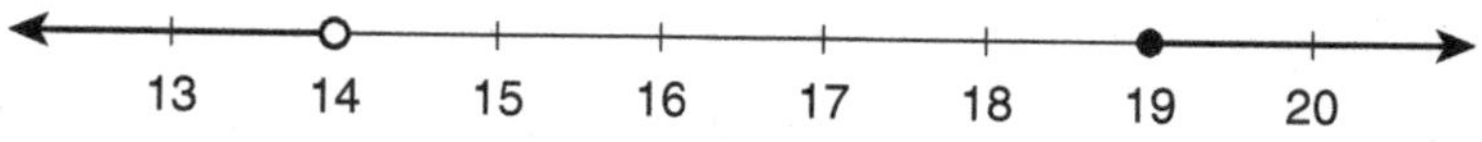

FIGURE 8.15 Graphical solution for Example 8.5: $x < 14$ or $x \geq 19$.

Example 8.6 Solve the compound inequalities and graph their solution set.

$$x + 11 \leq 15 \quad \text{or} \quad x + 1 \geq 13$$

Solution: ▶ *Treat each inequality expression separately. For example, subtract* **11** *from both sides of the inequality for the first expression and subtract* **1** *from the other expression.*

$$x + 11 - 11 \leq 15 - 11 \quad \text{or} \quad x + 1 - 1 \geq 13 - 1$$

$$x \leq 4 \quad \text{or} \quad x \geq 12. \qquad \blacktriangleleft$$

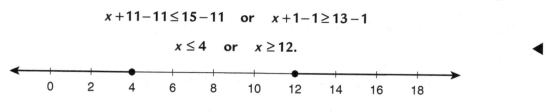

FIGURE 8.16 Graphical solution for Example 8.6: $x \leq 4$ or $x \geq 12$.

Example 8.7 Solve for **n** and graph the solution set.

$$n + 15 \leq 15 \quad \text{or} \quad n - 8 \geq -3$$

Solution: ▶ *Treat each inequality expression separately. For example, subtract **15** from both sides of the inequality for the first expression and add **8** to the other expression.*

$$n+15-15 \leq 15-15 \quad \text{or} \quad n-8+8 \geq -3+8$$

$$x \leq 0 \quad \text{or} \quad x \geq 5. \qquad \qquad \blacktriangleleft$$

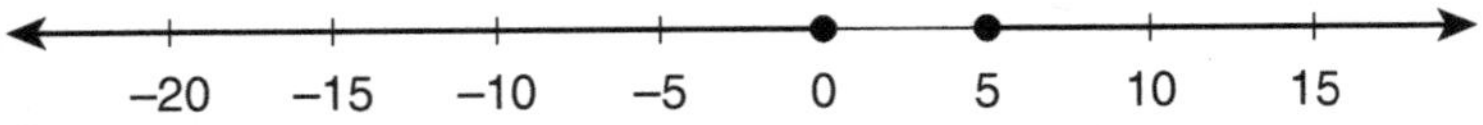

FIGURE 8.17 Graphical solution for Example 8.7: $x \leq 4$ or $x \geq 12$.

Supplementary Exercises

Exercise 8.4.1 *Find the set of all real numbers x satisfying the inequalities and graph solution set.*

$$x-10 > -15 \quad \text{or} \quad x+1 < -14,$$

Exercise 8.4.2 *Find the set of all real numbers x satisfying the inequalities and graph solution set.*

$$x-15 > -7 \quad \text{or} \quad x+19 < 13,$$

Exercise 8.4.3 *Find the set of all real numbers x satisfying the inequalities and graph solution set.*

$$16 \geq x+10 \geq 13,$$

Exercise 8.4.4 *Find the set of all real numbers x satisfying the inequalities and graph solution set.*

$$12 \geq x+17 \geq 4,$$

Exercise 8.4.5 *Find the set of all real numbers y satisfying the inequalities and graph solution set.*

$$-7 > y-17 > -9,$$

Exercise 8.4.6 *Find the set of all real numbers m satisfying the inequalities and graph solution set.*

$$-14 > m-19 > -17,$$

Exercise 8.4.7 *Find the set of all real numbers b satisfying the inequalities and graph solution set.*

$$b-4 \leq -5 \quad \text{or} \quad b+1 > 9,$$

Exercise 8.4.8 *Find the set of all real numbers a satisfying the inequalities and graph solution set.*

$$a-5 \geq 7 \quad \text{or} \quad a < 10,$$

Exercise 8.4.9 *Find the set of all real numbers z satisfying the inequalities and graph solution set.*

$$z-8 > -8 \quad \text{or} \quad z-12 < -22,$$

Exercise 8.4.10 *Find the set of all real numbers c satisfying the inequalities and graph solution set.*

$$-15 > c-13 > -17,$$

Exercise 8.4.11 *Find the set of all real numbers x satisfying the inequalities and graph solution set.*

$$x-1 \geq 15 \quad \text{or} \quad x+8 \leq 12,$$

Exercise 8.4.12 *Find the set of all real numbers y satisfying the inequalities and graph solution set.*

$$y+9 \geq 15 \quad \text{or} \quad y+4 < -2,$$

Exercise 8.4.13 *Find the set of all real numbers* x *satisfying the inequalities and graph solution set.*

$$-5 \le 2x + 1 \le 11,$$

Exercise 8.4.14 *Find the set of all real numbers* n *satisfying the inequalities and graph solution set.*

$$-3 \le n + 7 \le 17,$$

Exercise 8.4.15 *Find the set of all real numbers* p *satisfying the inequalities and graph solution set.*

$$17 > p + 16 > 11,$$

Exercise 8.4.16 *Find the set of all real numbers* x *satisfying the inequalities and graph solution set.*

$$x - 2 < 1 \quad \text{or} \quad x - 12 \ge -8,$$

Exercise 8.4.17 *Find the set of all real numbers* z *satisfying the inequalities and graph solution set.*

$$11 < z + 2 \le 17,$$

Exercise 8.4.18 *Find the set of all real numbers* p *satisfying the inequalities and graph solution set.*

$$-5 > p > -8.$$

Exercise 8.4.19 *Find the set of all real numbers* n *satisfying the inequalities and graph solution set.*

$$n - 9 > 7 \quad \text{or} \quad v - 20 < -6,$$

Exercise 8.4.20 *Find the set of all real numbers* v *satisfying the inequalities and graph solution set.*

$$v + 6 \le 7 \quad \text{or} \quad v + 6 > 14,$$

APPENDIX A:
REVIEW EXERCISES

CHAPTER 2 REVIEW EXERCISES

1. Which of the following expressions best represents: "the excess of a number over **3**"?

   (a) $\dfrac{x}{3}$

   (b) $x - 3$

   (c) $3x$

   (d) $\dfrac{3}{x}$

   (e) none of these

2. Which of the following expressions best represents: "the sum of a number plus thrice another being diminished by **3**"?

   (a) $\dfrac{x + 3y}{3}$

   (b) $x + 3y - 3$

   (c) $x + 3(y - 3)$

   (d) $x + 2y - 3$

   (e) none of these

3. The algebraic expression $x^2 - \dfrac{1}{x}$ can be translated as:

   (a) "The square of a number is increased by the reciprocal of the number."

   (b) "The square of a number is reduced by the reciprocal of the number."

   (c) "Twice a number is increased by the reciprocal of the number."

   (d) "Twice a number is decreased by the reciprocal of the number."

   (e) none of these

4. $\dfrac{5^3 - (-3)^3}{5^2 - (-3)^2} =$

   (a) $\dfrac{49}{2}$

   (b) $\dfrac{3}{2}$

   (c) 6

   (d) $\dfrac{19}{2}$

   (e) none of these

5. $(-100) \div (10) \div (-10) =$

   (a) 1

   (b) 10

   (c) 100

   (d) -1

   (e) none of these

6. $\dfrac{3}{4} - \dfrac{11}{12} =$

   (a) $\dfrac{1}{6}$ **(d)** $\dfrac{11}{16}$

   (b) $-\dfrac{1}{6}$ **(e)** none of these

   (c) -1

7. $\dfrac{3}{14} \div \dfrac{21}{8} =$

   (a) $\dfrac{9}{16}$ **(d)** $\dfrac{49}{4}$

   (b) $\dfrac{4}{49}$ **(e)** none of these

   (c) $\dfrac{24}{22}$

8. $\dfrac{(1 + 2 \cdot 3)^2 - 9}{(2 \cdot 5 - 2)(2^2 + 1)} =$

   (a) 1 **(d)** $\dfrac{2}{3}$

   (b) -1 **(e)** none of these

   (c) $\dfrac{29}{30}$

9. $1024 \div 512 \times 2 =$

   (a) 1 **(d)** 8

   (b) 2 **(e)** none of these

   (c) 4

10. $\dfrac{5^3 - 3^3}{5^2 + 15 + 3^2} =$

   (a) 1 **(d)** $\dfrac{8}{49}$

   (b) $\dfrac{1}{2}$ **(e)** none of these

   (c) 2

11. $1 - (-2)^3 =$

   (a) -7 **(d)** 7

   (b) 9 **(e)** none of these

   (c) -9

12. You have t \$10 and w \$20 bills. How much money, in dollars, do you have?

   (a) $t + w$ **(d)** $30 + t + w$

   (b) $10t + 20w$ **(e)** none of these

   (c) $200tw$

13. You have **t** $10 and **w** $ 20 bills. How many "paper bills" do you have?

(a) $t + w$

(b) $10t + 20w$

(c) $200tw$

(d) $30 + t + w$

(e) none of these

14. Consider the arithmetic progression **3, 14, 25, 36, ...**, where **3** is on the first position, **14** is on the second, etc. Which number occupies the **101**st position?

(a) **101**

(b) **1092**

(c) **1103**

(d) **1114**

(e) none of these

15. $\dfrac{1}{2} + \dfrac{2}{3} - \dfrac{6}{7} =$

(a) $\dfrac{13}{42}$

(b) -3

(c) $-\dfrac{1}{14}$

(d) $-\dfrac{53}{60}$

(e) none of these

16. $-5^2 + 4 - 9 - (-7) + (-9 + 5 \cdot 3)^2 =$

(a) -13

(b) -1

(c) 63

(d) 13

(e) none of these

17. Perform the calculation: $1 - 2 + 3 - 4 + 5 - 6 + 7 - 8 + 9 - 10$.

(a) **5**

(b) **0**

(c) **−5**

(d) **1**

(e) none of these

18. Perform the calculation: $\dfrac{\frac{3}{4}}{\frac{5}{6}}$

(a) $\dfrac{9}{10}$

(b) $\dfrac{10}{9}$

(c) $\dfrac{19}{12}$

(d) $\dfrac{5}{8}$

(e) none of these

19. Perform the calculation: $\dfrac{3}{4} + \dfrac{5}{6}$

(a) $\dfrac{9}{10}$

(b) $\dfrac{10}{9}$

(c) $\dfrac{19}{12}$

(d) $\dfrac{5}{8}$

(e) none of these

CHAPTER 3 REVIEW EXERCISES

20. Collect like terms: $(x + y + 2z) - (x - 2y + z)$.

    (a) $3y + z$

    (b) $2x - y + 3z$

    (c) $3y + 3z$

    (d) $3y - z$

    (e) none of these

21. Which of the following are like terms?

    (a) x^6 and $6x$

    (b) $2x^2$ and $-x^2$

    (c) j^2kl and jk^2l

    (d) c and c^2

    (e) none of these

22. $2(a - 2b) - 3(b - 2a) =$

    (a) $3a - b$

    (b) $-5a - 10b$

    (c) $8a - 7b$

    (d) $8a + 7b$

    (e) none of these

23. $\left(\dfrac{20x - 10}{5}\right) - \left(\dfrac{6 - 9x}{3}\right) =$

    (a) $7x$

    (b) $7x - 4$

    (c) x

    (d) $x - 4$

    (e) none of these

24. Collect like terms: $2(a - 2b) - (2a - b)$

    (a) $-3b$

    (b) $-b$

    (c) $4a - 5b$

    (d) $4a - 3b$

    (e) none of these

25. Collect like terms: $-2(x^2 - 1) + 4(2 - x - x^2)$

    (a) $-6x^2 - 4x + 10$

    (b) $6x^2 + 4x + 10$

    (c) $-6x^2 + 4x + 10$

    (d) $-2x^2 - 4x + 10$

    (e) none of these

26. Collect like terms: $(-2a + 3b - c) + (5a - 8b + c) =$

    (a) $7a - 11b - 2c$

    (b) $3a - 5b - 2c$

    (c) $-7a + 11b - 2c$

    (d) $3a - 5b$

    (e) none of these

27. $-a + 2b + 3a - 4b =$

    (a) $-2a + 2b$

    (b) $2a - 2b$

    (c) $-4a - 6b$

    (d) $2a + 2b$

    (e) none of these

28. $\dfrac{a}{3} - \dfrac{a}{5} =$

    (a) $\dfrac{a}{15}$

    (b) $-\dfrac{a}{2}$

    (c) $\dfrac{2a}{15}$

    (d) $\dfrac{a}{8}$

    (e) none of these

29. $\dfrac{a}{3} + \dfrac{a^2}{3} + \dfrac{a}{4} + \dfrac{3a^2}{4} =$

    (a) $\dfrac{4a}{7} + \dfrac{4a^2}{7}$

    (b) $\dfrac{7a}{12} + \dfrac{13a^2}{12}$

    (c) $\dfrac{a}{12} + \dfrac{a^2}{12}$

    (d) $\dfrac{a^6}{48}$

    (e) none of these

30. $(-a + 2b - 3c) - (-a + 2b + 3c) =$

    (a) $-6c$

    (b) $4b - 6c$

    (c) $2a + 4b + 6c$

    (d) $2a - 6c$

    (e) none of these

31. Collect like terms: $\dfrac{3x-6}{3} + x - 1.$

    (a) $2x - 2$

    (b) $2x - 3$

    (c) $x - 3$

    (d) $2A - B - 2$

    (e) none of these

32. How many of the following expressions are equivalent to $5 + x^2 y$?

    I: $x^2 y + 5$,   *II*: $5 + yx^2$,   *III*: $5 + xy^2$,   *IV*: $yx^2 + 5$.

    (a) exactly one

    (b) exactly two

    (c) exactly three

    (d) all four

    (e) none

33. Collect like terms: $\left(\dfrac{-35x + 20}{5} \right) + \left(\dfrac{12 - 24x}{6} \right).$

    (a) $-11x + 6$

    (b) $3x + 2$

    (c) $3x + 6$

    (d) $-3x + 2$

    (e) none of these

CHAPTER 4 REVIEW EXERCISES

34. $\dfrac{3^4}{3^3} =$

    (a) 3

    (b) 3^{12}

    (c) $\dfrac{1}{3}$

    (d) 3^7

    (e) none of these

35. $\dfrac{3^3}{3^4} =$

    (a) 3

    (b) 3^{12}

    (c) $\dfrac{1}{3}$

    (d) 3^7

    (e) none of these

36. $(3^3)^4 =$

    (a) 3

    (b) 3^{12}

    (c) $\dfrac{1}{3}$

    (d) 3^7

    (e) none of these

37. $3^3 \cdot 3^4 =$

    (a) 3

    (b) 3^{12}

    (c) $\dfrac{1}{3}$

    (d) 3^7

    (e) none of these

38. $\dfrac{3^3}{3^{-4}} =$

    (a) 3

    (b) 3^{12}

    (c) $\dfrac{1}{3}$

    (d) 3^7

    (e) none of these

39. $\dfrac{1111^5 + 1111^5 + 1111^5 + 1111^5}{1111^4 + 1111^4} =$

    (a) 1111

    (b) 2222

    (c) 4444

    (d) $\dfrac{1}{2222}$

    (e) none of these

40. Simplify and write with positive exponents only: $(ab^2)^2(a^3b^{-4})$.

(a) a^5

(b) a^5b^8

(c) $\dfrac{b^8}{a}$

(d) a^6

(e) none of these

41. Simplify and write with positive exponents only: $(ab^2)^2 \div (a^3b^{-4})$.

(a) a^5

(b) a^5b^8

(c) $\dfrac{b^8}{a}$

(d) a^6

(e) none of these

42. Perform the calculation: $\dfrac{2^{-3}}{3^{-2}} - \dfrac{3^2}{2^2}$.

(a) $\dfrac{9}{8}$

(b) $\dfrac{9}{4}$

(c) $-\dfrac{9}{8}$

(d) $-\dfrac{9}{4}$

(e) none of these

43. Perform the calculation: $\dfrac{1}{2^{-5}} - 5^{-2}$.

(a) 0

(b) 20

(c) $\dfrac{799}{25}$

(d) $\dfrac{2}{25}$

(e) none of these

44. Multiply and write with positive exponents only: $(x^4y^7)(x^{-2}y^{-8})$.

(a) x^6y^{15}

(b) x^2y

(c) $\dfrac{x^2}{y}$

(d) $\dfrac{y}{x^2}$

(e) none of these

45. Multiply and collect like terms: $(x-2)(x^2+2x+4)$.

(a) $x^3 - 8$

(b) $x^3 - 4x^2 - 4x - 8$

(c) $x^3 + 8$

(d) $x^3 - 4x^2 - 8$

(e) none of these

46. Divide and write with positive exponents only $\dfrac{x^4y^7}{x^{-2}y^{-8}}$.

(a) x^6y^{15}

(b) x^2y

(c) $\dfrac{x^2}{y}$

(d) $\dfrac{y}{x^2}$

(e) none of these

47. $\dfrac{2^{-3}}{3^{-2}2^3} =$

    (a) $\dfrac{9}{64}$

    (b) 9

    (c) $\dfrac{64}{9}$

    (d) $\dfrac{1}{9}$

    (e) none of these

48. $(x^{-1}y^{-2}z^3)^{-3} =$

    (a) $x^4y^5z^0$

    (b) $\dfrac{z^9}{x^3y^6}$

    (c) $\dfrac{x^3y^6}{z^9}$

    (d) $\dfrac{x^3y^9}{y^6}$

    (e) none of these

49. $(x^4\,y^3\,z^2)\,(x^2\,y^4\,z^2) =$

    (a) $\dfrac{x^2}{y}$

    (b) x^2y

    (c) $\dfrac{y}{x^2}$

    (d) $x^{\,6}y^7\,z^4$

    (e) none of these

50. Multiply and collect like terms: $(x+2)(x-3) =$

    (a) $x^2 - 6$

    (b) $x^2 - x - 6$

    (c) $x^2 + x - 6$

    (d) $x^2 - 5x - 6$

    (e) none of these

51. Multiply and collect like terms: $(2x - 1)(2x + 1) - 2(2x - 1) =$

    (a) $4x^2 - 4x$

    (b) $4x^2 - 4x + 1$

    (c) $4x^2 - 4x - 1$

    (d) 1

    (e) none of these

52. Multiply and collect like terms: $(a + 2)\,(a - 2) - (a - 1)\,(a + 1) =$

    (a) $2a^2 - 3$

    (b) -3

    (c) -5

    (d) $2a^2 - 5$

    (e) none of these

53. Multiply and collect like terms: $(a + 2)^2 + (a - 2)^2 =$

    (a) $2a^2 + 8$

    (b) $8a$

    (c) $2a^2 + 4$

    (d) $2a^2 - 4a + 8$

    (e) none of these

54. Multiply and collect like terms: $(a+2)^2 - (a-2)^2 =$

   (a) $2a^2 + 8$

   (b) $8a$

   (c) $2a^2 + 4$

   (d) $2a^2 - 4a + 8$

   (e) none of these

55. Expand and collect like terms: $(x+2)(x^2 - 4x + 1)$

   (a) $x^3 - 6x^2 + 9x - 2$

   (b) $x^3 - 2x^2 - 7x + 2$

   (c) $x^3 + 2x^2 - 9x + 2$

   (d) $x^4 - 1$

   (e) none of these

CHAPTER 5 REVIEW EXERCISES

56. Perform the division: $(x^3 - 8x^2) \div (-x)$.

   (a) $x^2 - 8x$

   (b) $x^2 + 8x$

   (c) $-x^2 + 8x$

   (d) $-x^2 - 8x$

   (e) none of these

57. Factor: $2x^2 - 3x$.

   (a) $x(2x - 3)$

   (b) $2x(x - 3)$

   (c) $x^2(2x - 3)$

   (d) $x(2x + 3)$

   (e) none of these

58. Factor: $x^2 + 3x + 2$.

   (a) $(x + 1)(x - 2)$

   (b) $(x + 1)(x + 2)$

   (c) $(x - 1)(x + 2)$

   (d) $(x - 1)(x - 2)$

   (e) none of these

59. Factor: $x^2 - 3x + 2$.

   (a) $(x + 1)(x - 2)$

   (b) $(x + 1)(x + 2)$

   (c) $(x - 1)(x + 2)$

   (d) $(x - 1)(x - 2)$

   (e) none of these

60. Factor: $x^2 - x - 2$.

   (a) $(x + 1)(x - 2)$

   (b) $(x + 1)(x + 2)$

   (c) $(x - 1)(x + 2)$

   (d) $(x - 1)(x - 2)$

   (e) none of these

61. $\dfrac{2x^3 - 3x^2 - 4x - 1}{2x + 1} =$

   (a) $x^2 + 2x + 1$

   (b) $x^2 - 2x - 1$

   (c) $x^2 - 2x + 1$

   (d) $x^2 + 2x - 1$

   (e) none of these

62. $\dfrac{6x^2 + 2x}{2x} =$

    (a) $3x + 1$ (d) $6x^2$

    (b) $3x$ (e) none of these

    (c) $3x + 2$

63. $\dfrac{(6x^2)(2x)}{2x} =$

    (a) $3x + 1$ (d) $6x^2$

    (b) $3x$ (e) none of these

    (c) $3x + 2$

64. $(9a^3b^3 - 6a^2b^4 + 3a^2b^3) \div (3a^2b^3)$

    (a) $3a - 2b$ (d) $3a^5b^6 - 2a^4b^7 + a^4b^6$

    (b) $3a - 2b + 1$ (e) none of these

    (c) $-6ab$

65. $(x^2 + 5x - 14) \div (x - 2) =$

    (a) $x - 7$ (d) $x + 12$

    (b) $x + 7$ (e) none of these

    (c) $x + 2$

66. $(6x^3 - 5x^2 + 7x - 2) \div (3x - 1) =$

    (a) $2x^2 + x + 2$ (d) $2x^2 + x - 2$

    (b) $2x^2 - x - 2$ (e) none of these

    (c) $2x^2 - x + 2$

67. $(x^3 + 27) \div (x + 3) =$

    (a) $x^2 - 9x + 9$ (d) $x^2 - 3x + 9$

    (b) $x^2 + 9x + 9$ (e) none of these

    (c) $x^2 + 3x + 9$

68. $\dfrac{x^9 + x^6}{x^3} =$

    (a) $x^3 + x^2$ (d) x^{12}

    (b) $3x^3 + 2x^2$ (e) none of these

    (c) $x^6 + x^3$

69. $\dfrac{(x^9)(x^6)}{x^3} =$

    (a) $x^3 + x^2$ (d) x^{12}

    (b) $3x^3 + 2x^2$ (e) none of these

    (c) $x^6 + x^3$

70. Factor: $xy^2 - x^2y$.

   (a) $xy(x - y)$

   (b) $xy(y - x)$

   (c) $x^2y^2(x - y)$

   (d) $xy(x + y)$

   (e) none of these

71. Factor: $x^2 - x - 12$.

   (a) $(x-3)(x + 4)$

   (b) $(x-6)(x + 2)$

   (c) $(x + 3)(x-4)$

   (d) $(x- 12)(x + 1)$

   (e) none of these

72. Factor: $x^2 - 12x + 27$.

   (a) $(x - 3)(x - 9)$

   (b) $(x + 3)(x + 9)$

   (c) $(x-3)(x + 9)$

   (d) $(x- 1)(x-27)$

   (e) none of these

73. Factor: $x^3 - 12x^2 + 27x$.

   (a) $x(x - 3)(x - 9)$

   (b) $x(x + 3)(x + 9)$

   (c) $x(x - 3)(x + 9)$

   (d) $x(x - 1)(x - 27)$

   (e) none of these

74. Factor: $a^2 + 16a + 64$.

   (a) $(a + 8)^2$

   (b) $(a - 8)(a + 8)$

   (c) $(a - 8)^2$

   (d) $(a + 16)(a + 4)$

   (e) none of these

75. Factor: $(a^3 + 16a^2 + 64a)$.

   (a) $a(a + 8)^2$

   (b) $a(a - 8)(a + 8)$

   (c) $a(a - 8)^2$

   (d) $a(a + 16)(a + 4)$

   (e) none of these

76. Factor: $a^2 - 64$.

   (a) $(a + 8)^2$

   (b) $(a - 8)(a + 8)$

   (c) $(a - 8)^2$

   (d) $(a +16)(a + 4)$

   (e) none of these

77. Divide: $(6x^3 + x^2 - 11x - 6) \div (3x + 2)$

   (a) $2x^2 - x + 3$

   (b) $2x^2 - x - 3$

   (c) $2x^2 + x - 3$

   (d) $2x^2 + x + 3$

   (e) none of these

78. Reduce the fraction: $\frac{x^3 - x}{x^2 - 1}$

   (a) x

   (b) $\frac{1}{x}$

   (c) $x + 1$

   (d) $x - 1$

   (e) none of these

79. Add: $\frac{2}{x-2} + \frac{3}{x+3}$

   (a) $\frac{5x}{x^2 + x + 6}$

   (b) $\frac{5x}{x^2 + x - 6}$

   (c) $\frac{2x + 1}{x^2 + x - 6}$

   (d) $\frac{12 - x}{x^2 + x - 6}$

   (e) none of these

80. Reduce the fraction: $\frac{a^2 - a}{a^2 - a}$.

   (a) $\frac{a^2 - 1}{a^2 + 1}$

   (b) -1

   (c) $\frac{a - 1}{a + 1}$

   (d) $\frac{a}{a - 1}$

   (e) none of these

81. Reduce the fraction: $\frac{a^2}{a^2 + a}$.

   (a) $\frac{a}{a + 1}$

   (b) 1

   (c) $\frac{1}{a + 1}$

   (d) $\frac{a}{a - 1}$

   (e) none of these

82. Add the fractions: $\frac{2}{x-3} + \frac{3}{x+2}$.

   (a) $\frac{5x - 5}{x^2 - x - 6}$

   (b) $\frac{13 + x}{x^2 + x - 6}$

   (c) $\frac{13 - x}{x^2 - x - 6}$

   (d) $\frac{5x + 5}{x^2 + x - 6}$

   (e) none of these

83. Perform the subtraction: $\frac{2}{x-3} - \frac{3}{x+2}$.

(a) $\frac{5x-5}{x^2-x-6}$

(b) $\frac{13+x}{x^2+x-6}$

(c) $\frac{13-x}{x^2-x-6}$

(d) $\frac{5x+5}{x^2+x-6}$

(e) none of these

CHAPTER 6 REVIEW EXERCISES

84. Solve the following equation for x: $ax = 2$.

(a) $x = \frac{2}{a}$

(b) $x = 2 - a$

(c) $x = \frac{a}{2}$

(d) $x = a$

(e) none of these

85. Solve the following equation for x: $\frac{ax}{b} = 2a$.

(a) $x = b$

(b) $x = \frac{2a}{2}$

(c) $x = 2b$

(d) $x = \frac{2}{b}$

(e) none of these

86. Solve the following equation for x: $ax + 2b = 5b$.

(a) $x = 7ab$

(b) $x = \frac{7b}{a}$

(c) $x = 3ab$

(d) $x = \frac{3b}{a}$

(e) none of these

87. Solve the following equation for x: $2x - 1 = x$.

(a) $x = 1$

(b) $x = -1$

(c) $x = \frac{1}{2}$

(d) $x = \frac{1}{3}$

(e) none of these

88. Solve the following equation for x: $1 - x = 2$.

(a) $x = 3$

(b) $x = -1$

(c) $x = -3$

(d) $x = \frac{1}{3}$

(e) none of these

89. Solve the following equation for x: $2x - 1 = 1 - 3x$.

(a) $x = \dfrac{5}{2}$

(b) $x = \dfrac{2}{5}$

(c) $x = 0$

(d) $x = -\dfrac{2}{5}$

(e) none of these

90. Solve the following equation for x: $2(x - 1) = 3(1 - 3x)$.

(a) $x = \dfrac{5}{11}$

(b) $x = \dfrac{1}{7}$

(c) $x = \dfrac{5}{7}$

(d) $x = -\dfrac{11}{5}$

(e) none of these

91. Solve the following equation for x: $\dfrac{x-1}{3} = \dfrac{1-3x}{2}$.

(a) $x = \dfrac{5}{11}$

(b) $x = \dfrac{1}{7}$

(c) $x = \dfrac{5}{7}$

(d) $x = -\dfrac{11}{5}$

(e) none of these

92. Solve the following equation for x: $ax - 2b = 5b$.

(a) $x = 7ab$

(b) $x = \dfrac{7b}{a}$

(c) $x = 3ab$

(d) $x = -\dfrac{3b}{a}$

(e) none of these

93. Solve the following equation for x: $\dfrac{x}{2} - \dfrac{3}{4} = \dfrac{9}{8}$.

(a) $x = \dfrac{3}{4}$

(b) $x = \dfrac{1}{8}$

(c) $x = \dfrac{15}{8}$

(d) $x = \dfrac{15}{4}$

(e) none of these

94. Adam, Eve, Snake, and Jake share **$240** between them. Adam has twice as much as Eve, Eve has twice as much as Snake, and Snake has twice as much as Jake. How much money does Eve have?

(a) **$15**

(b) **$60**

(c) **$64**

(d) **$128**

(e) none of these

95. Peter, Paul, and Mary share **$153** between them. Peter has twice as much as Paul, and Mary has thrice as much as Peter. How much money does Peter have?

    (a) $17

    (b) $34

    (c) $51

    (d) $102

    (e) none of these

96. The sum of three consecutive odd integers is **909**. Which one is the last number?

    (a) 301

    (b) 302

    (c) 303

    (d) 305

    (e) none of these

97. Jane's age is twice Bob's age increased by **3**. Bill's age is Bob's age decreased by **4**. If the sum of their ages is **27**, how old is Bill?

    (a) 3

    (b) 7

    (c) 4

    (d) 17

    (e) none of these

98. A jar of change contains nickels, dimes, and quarters. There are twice as many dimes as nickels, and twice as many quarters as dimes. If the jar has a total of $8.75, how many quarters are there?

    (a) 5

    (b) 7

    (c) 14

    (d) 28

    (e) none of these

99. Currently, the age of a father is four times the age of his son, but in **24** years from now the father's age will only be double his son's age. Find the son's current age.

    (a) 12

    (b) 36

    (c) 48

    (d) 6

    (e) none of these

CHAPTER 7 REVIEW EXERCISES

100. Solve for x: $2x^2 - 3x = 0$.

    (a) $x = 0$ or $x = \dfrac{3}{2}$

    (b) $x = 0$ or $x = \dfrac{2}{3}$

    (c) $x = 0$ or $x = -\dfrac{3}{2}$

    (d) $x = 0$ or $x = -\dfrac{2}{3}$

    (e) none of these

101. Solve for x: $x^2 + 3x + 2 = 0$.

    (a) $x = -1$ or $x = 2$

    (b) $x = -1$ or $x = -2$

    (c) $x = 1$ or $x = -2$

    (d) $x = 1$ or $x = 2$

    (e) none of these

102. Solve for x: $x^2 - 3x + 2 = 0$.

    (a) $x = -1$ or $x = 2$

    (b) $x = -1$ or $x = -2$

    (c) $x = 1$ or $x = -2$

    (d) $x = 1$ or $x = 2$

    (e) none of these

103. Solve for x: $x^2 - x - 2 = 0$.

    (a) $x = -1$ or $x = 2$

    (b) $x = -1$ or $x = -2$

    (c) $x = 1$ or $x = -2$

    (d) $x = 1$ or $x = 2$

    (e) none of these

104. If $2(x - 1) \geq x + 3$ then

    (a) $x \geq 1$

    (b) $x \geq 5$

    (c) $x \leq 1$

    (d) $x \leq 5$

    (e) none of these

105. If $-2x > x + 3$ then

    (a) $x > -1$

    (b) $x > 1$

    (c) $x < -1$

    (d) $x < 1$

    (e) none of these

106. Which graph gives the correct solution to $2 - 5x \leq -3$?

    (a)

    (b)

    (c)

    (d)

    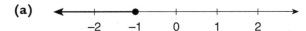

    (e) none of these

107. Which graph gives the correct solution to $2 + 5x \leq -3$?

    (a)

    (b)

    (c)

    (d)

    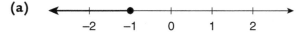

    (e) none of these

True or False Exercises.

108. ☐ $(2+3)^{10} = 2^{10} + 3^{10}$.

109. ☐ $(2 \cdot 3)^{10} = 2^{10} \cdot 3^{10}$.

110. ☐ $(2^5)^2 = 2^7$.

111. ☐ It is always true that $x(2x+3) = x(2x)(3)$.

112. ☐ $(1^1 + 2^2 + 3^3 + 4^4 - 5^5)0 + 4 = 5$.

In each item below, select the choice(s) that indicates equivalence to the given expression. Denominators are presumed not to vanish.

113. $-a =$

    (a) $a(-1)$

    (b) $-1 \cdot a$

    (c) $\dfrac{a}{-1}$

    (d) $\dfrac{-a}{1}$

    (e) $1 - 2a$

114. $x\,y - z =$

    (a) $z - x\,y$

    (b) $yx - z$

    (c) $x\,y + (-z)$

    (d) $x(y - z)$

    (e) $-z + x\,y$

115. $\dfrac{a-b}{2} =$

    (a) $a - b \div 2$

    (b) $(a - b) \div 2$

    (c) $\dfrac{a}{2} - \dfrac{b}{2}$

    (d) $\dfrac{-b+a}{2}$

    (e) $\dfrac{1}{2}(a - b)$

116. $10x\left(\dfrac{y}{5} + x\right) =$

    (a) $\dfrac{xy}{2} + 10x^2$

    (b) $2xy + 10x$

    (c) $10\left(\dfrac{Y}{5} + x\right)x$

    (d) $5x \cdot \left(x + \dfrac{y}{5}\right) \cdot 2$

    (e) $2xy + 10x^2$

117. $(a - 2b)^2 =$

    (a) $(a - 2b)(a - 2b)$

    (b) $a^2 - 4b^2$

    (c) $a^2 + 4b^2$

    (d) $a^2 - 4ab + 4b^2$

    (e) $-(2b - a)(a - 2b)$

118. $\dfrac{x-y}{y-x} =$

   (a) 1

   (b) −1

   (c) −2

   (d) $(x-y)^2$

   (e) $\dfrac{y-x}{x-y}$

119. $(2x^2y^2)(3x\,y^4) =$

   (a) $6x^3y^6$

   (b) $6(x\,y^4)^2$

   (c) $6(x\,y^2)^3$

   (d) $\dfrac{12x^9y^{12}}{2x^3y^2}$

   (e) $\dfrac{12x^6y^7}{2x^3y}$

120. $x(x^3-1)+2(x^3-1) =$

   (a) $(x^3-1)x -2(-x^3+1)$

   (b) $2(x^3-1)+x(-x^3+1)$

   (c) $(x+2)(x^3-1)$

   (d) x^4-x+2x^3-2

   (e) $x^4-x+1(x^3-2)$

CHAPTER 8 REVIEW EXERCISES

121. $3x-4 \le 2 =$

   (a) $x \le 2$

   (b) $x \le 3$

   (c) $x \ge 2$

   (d) $x \ge -\dfrac{2}{3}$

   (e) none of these

122. $\dfrac{3x}{5} > 15 =$

   (a) $x < 25$

   (b) $x > 5$

   (c) $x > 25$

   (d) $x > 1$

   (e) none of these

123. $\dfrac{10-2x}{7} \ge 20 =$

   (a) $x \ge 130$

   (b) $x \ge 65$

   (c) $x > 25$

   (d) $x > 1$

   (e) none of these

124. $3(m-2) < m =$

   (a) $m > 3$

   (b) $m < 3$

   (c) $m < 6$

   (d) $m > 6$

   (e) none of these

125. $-28 < 3x - 4 < -10 =$

    (a) $-8 < x < -2$

    (b) $8 < x < 2$

    (c) $x > -8$

    (d) $x < -2$

    (e) none of these

126. $25 \le 4y - 3 \le 9 =$

    (a) $7 < y < 3$

    (b) $7 \le y \le 3$

    (c) $3 \le y \le 7$

    (d) $-7 \le y \le -3$

    (e) none of these

127. $8(n-1) + 5 \ge 3(5n - 4) =$

    (a) $n \le \dfrac{7}{9}$

    (b) $n \le \dfrac{9}{7}$

    (c) $n \le 7$

    (d) $n \le 9$

    (e) none of these

128. $(n-1)(n-3) \le 4n =$

    (a) $n \le \dfrac{3}{4}$

    (b) $n \ge \dfrac{3}{4}$

    (c) $n^2 \le \dfrac{3}{4}$

    (d) $n^2 \ge \dfrac{3}{4}$

    (e) none of these

APPENDIX B:
ANSWER KEYS FOR CHAPTER SUPPLEMENTARY EXERCISES

CHAPTER 1 ANSWERS

1.2.1 The Alexandrian mathematician Diophantus, for the introduction of syncopation (symbols to express abstract quantities and procedures on them) in algebra.

1.2.2 The word *algebra* is Arabic, but algebra was studied long before the Arabs had any prominent role in history.

1.2.3 The word *ossifrage* is Latin for "breaking bones." The word *algebra* is Arabic for "bone setting."

1.3.1 You need to increase **$20**, and **$20** is **25%** of **$80**, and so you need to increase by **25%**.

1.3.2 **10^c**. The cork costs **10^c** and the bottle **90^c**.

1.4.1 St. Augustine is using the word *mathematician* in lieu of the word *astrologer*.

1.4.2 No. When we add an even integer to another even integer the result is an even integer. Thus the sum of five even integers is even, but **25** is odd.

1.4.3 The **27**th floor.

1.4.4 Between **9999** and **10101**.

1.4.5 $\boxed{2}$ 75 $\boxed{3}$ 6 – $\boxed{2}$ 56 $\boxed{1}$ = 24 $\boxed{9}$ 75.

1.4.6 Observe that the first and second rows, and the second and third columns, add up to **8**. Thus *a* = **4** works.

1.4.7 There are multiple solutions. They can be obtained by permuting the entries of one another. Here are two:

8	1	6
3	5	7
4	9	2

4	9	2
3	5	7
8	1	6

1.4.8 We can represent the brother's and sister's amounts by boxes, each box having an equal amount of coins, where two boxes are for the brother and one for the sister:

$\boxed{B}$ $\boxed{B}$ $\boxed{G}$

Then clearly, each box must have **8** coins. The brother has **16** and the sister has **8**.

CHAPTER 2 ANSWERS

2.1.1 $N - 20$.

2.1.2 $N + 20$.

2.1.3 $3x + 10$.

2.1.4 $3(x + 10)$.

2.1.5 In crocheting **s** baby blankets, I need **5s** balls of yarn. Therefore, at the end of the day I have **b − 5s** balls of yarn.

2.1.6 On the first step you have **x**. On the second step you have **2x**. On the third step you have **2x + 10**. On the fourth step you have $\frac{2x+10}{2}$. On the fifth step you have $\frac{2x+10}{2} - x$. You are asserting that $\frac{2x+10}{2} - x$ is identically equal to **5**.

2.1.7 **4n; 4n + 1; 4n + 2; 4n + 3**, where **n** is a natural number

2.1.8 **6n + 1**, where **n = 0, 1, 2, 3, …**.

2.1.9 First observe that

$$3 = 2 \times 1 + 1,$$

$$5 = 2 \times 2 + 1,$$

$$7 = 2 \times 3 + 1,$$

etc. Then it becomes clear that if the last number on the left is **n**, then the right-hand side is **n(n + 1)(2n + 1)** divided by **6**. Therefore we are asserting that

$$1^2 + 2^2 + 3^2 + \cdots (n-1)^2 + n^2 = \frac{(n)(n+1)(2n+1)}{6}.$$

2.1.10 The general formula is

$$1 + 3 + \ldots + (2n - 1) = n^2.$$

2.1.11 You have **25q + 10d** cents at the beginning of the day, and then you lose **25a +10b**, which leaves you with

$$(25q + 10d) - (25a + 10b)$$

cents.

2.1.12 The cube is x^3, the square is x^2. The cube reduced by the square is $x^3 - x^2$. In conclusion, what is left after division by **8** is

$$\frac{x^3 - x^2}{8}.$$

2.1.13 The jacket costs **3h** and the jeans cost **3h − 8**. Therefore he paid in total

$$h + 3h + 3h - 8 = 7h - 8.$$

2.1.14 The profit is that amount over the real price. If the real price is Γ, then $\Gamma + b$ is the real price plus the profit, that is, the farmer's selling price for the cow. In other words, $\Gamma + b = a$, therefore $\Gamma = a - b$.

2.1.15 The salesman starts with **s** and gains **x** for a total of **s + x**. He then loses **y**, for a total of **s + x − y**. After he gains **b** more dollars he has a running total of **s + x − y + b**, and finally, after losing **z** dollars he has a grand total of **s + x − y + b − z**.

2.2.1 108.

2.2.2 1.

2.2.3 28.

2.2.4 21.

2.2.5 15.

2.2.6 Since $100 = 7 \times 14 + 2$, **98** days from today will be a Thursday, and **100** days from today will be a Saturday.

2.2.7 Observe that **4** and **5** are on opposite sides, so they never meet. The largest product is $3 \times 5 \times 6 = 90$.

2.2.8 The first contractor prints $4500 \div 30 = 150$ cards per day, and the second, $4500 \div 45 = 100$ cards per day. Therefore, if they worked simultaneously, they could print $150 + 100 = 250$ cards per day, and it would take them $4500 \div 250 = 18$ days to print all cards.

2.2.9 Seven.

2.2.10 9 minutes.

2.2.11 If $36 \div n$ is a natural number, then n must evenly divide **36**, which means that n is a divisor of **36**. Thus n can be any one of the values in $\{1, 2, 3, 4, 6, 9, 12, 18, 36\}$.

2.2.12 From the distributive law,

$$
\begin{aligned}
(666)(222) + (1)(333) + (333)(222) & \\
+ (666)(333) + (1)(445) + (333)(333) & \\
+ (666)(445) + (333)(445) + (1)(222) &= (666 + 333 + 1)(445 + 333 + 222) \\
&= (1000)(1000) \\
&= 1000000.
\end{aligned}
$$

2.2.13 Travelling **30,000** miles with **4** tires is equivalent to travelling **120,000** miles on one tire. The average wear of each of the **5** tires is thus $120000 \div 5 = 24000$ miles.

2.2.14 Anna answered **20** questions correctly. (She could not answer less than **20**, because then her score would have been less than $19 \times 4 = 76 < 77$, and she could not answer more than **20**, because her score would have been at least $21 \times 4 - 3 = 81.77$). To get exactly **77** points, Anna had to answer exactly **3** questions wrong, which means she omitted **2** questions.

2.2.15 The trick is to use a technique analogous to the one for multiplying, but this time using three digits at a time:

$$321 \times 654 = 209934,$$

$$321 \times 987 = 316827,$$

$$745 \times 654 = 487230,$$

$$745 \times 987 = 735315.$$

Thus

$$
\begin{array}{r}
\boxed{987}\ \ \boxed{654} \\
\times \qquad \boxed{745}\ \ \boxed{321} \\
\hline
209\ \ 934 \\
316\ \ 827 \qquad\quad \\
735\ \ 315 \qquad\qquad\quad \\
\hline
736\ \ 119\ \ 266\ \ 934
\end{array}
$$

2.2.16 There are **28** digits, since

$$4^{16}5^{25} = 2^{32}5^{25} = 2^{7}2^{25}5^{25} = 128 \times 10^{25},$$

which is the **3** digits of **128** followed by **25** zeroes.

2.2.17 The least common multiple of **4, 5,** and **6** is **60**, therefore the smallest positive multiple of **60** leaving remainder **1** upon division by **7**. This is **120**.

2.2.18 Using the definition of $\oplus$,

$$1 \oplus (2 \oplus 3) = 1 \oplus (1+2(3)) = 1 \oplus 7 = 1 + 1(7) = 8,$$

but

$$(1 \oplus 2) \oplus 3 = (1+1(2)) \oplus 3 = 3 \oplus 3 = 1 + 3(3) = 10.$$

Since these two answers differ, the operation is not associative.

Now, for the operation to be commutative, we must have, in every instance, $a \oplus b = b \oplus a$. Using the fact that multiplication of numbers is commutative, we deduce

$$a \oplus b = 1 + ab = 1 + ba = b \oplus a,$$

so our operation is indeed commutative.

2.3.1 Number the steps from **0** to N, where N is the last step, and therefore there are $N + 1$ steps. Notice that the top and the bottom of the stairs are counted as steps. Let us determine first the number of steps. Freddy steps on steps **0, 2, 4,** … , N; Lorance steps on steps **0, 3, 6,** … , N; George steps on steps **0, 4, 8,** … , N; and Aragorn steps on steps **0, 5, 10,** … , N. For this last step to be common, it must be a common multiple of **2, 3, 4,** and **5**, and therefore

$$N = \mathsf{lcm}(2, 3, 4, 5) = 60.$$

Since every step that George covers is also covered by Freddy, we don't consider George in our count. Freddy steps alone on the steps which are multiples of **2**, but not multiples of **3** or **5**. Therefore he steps on the **8** steps

$$\textbf{2, 14, 22, 26, 34, 38, 46, 58.}$$

Lorance steps alone on the **8** steps that are multiples of **3**, but not multiples of **2** nor **5**:

$$\textbf{3, 9, 21, 27, 33, 39, 51, 57.}$$

Finally, Steve steps alone on the **4** steps that are multiples of **5** but not multiples of **2** or **3**:

$$\textbf{5, 25, 35, 55.}$$

Our final count is thus $8 + 8 + 4 = 20$.

2.3.2 $\frac{36}{60} = \frac{3}{5}$ of an hour.

2.3.3 $\frac{17}{35}$.

2.3.4 Observe that $3990 = 210 \cdot 19$. Thus $\frac{102}{210} = \frac{102 \cdot 19}{210 \cdot 19} = \frac{1938}{3990}$.

2.3.5 Observe that $3 \cdot 5 \cdot 13 = 195$ is a common denominator. Now

$$\frac{2}{3} = \frac{130}{195}; \frac{3}{5} = \frac{117}{195}; \frac{8}{13} = \frac{120}{195},$$

therefore $\frac{3}{5}$ is the smallest, $\frac{8}{13}$ is in the middle, and $\frac{2}{3}$ is the largest.

2.3.6 Three lines have been sung before the fourth singer starts, and after that he sings 12 more lines. So the total span of lines is 15. They start singing simultaneously from line 4, and the first singer is the first to end, in line 12. Thus $12 - 4 + 1 = 9$ lines out of 15 are sung simultaneously and the fraction sought is $\frac{9}{15} = \frac{3}{5}$.

2.4.1 Here are the solutions.

$\frac{1}{3}$	+	$\frac{5}{3}$	=	2
+	■	−	■	
$\frac{3}{4}$	×	1	=	$\frac{3}{4}$
=	■	=	■	
$\frac{13}{12}$	÷	$\frac{2}{3}$	=	$\frac{13}{8}$

2.4.2 $\frac{7}{45}$.

2.4.3 $\frac{3}{17}$.

2.4.4 We have,

$$\frac{10 + 10^2}{\frac{1}{10} + \frac{1}{100}} = \frac{10^3 + 10^4}{10 + 1} = \frac{11000}{11} = 1000.$$

2.4.5 1.

2.4.6 We have,

$$\frac{1}{1 + \frac{1}{5}} = \frac{1}{\frac{6}{5}} = \frac{5}{6} = \frac{a}{b},$$

therefore $a^2 + b^2 = 5^2 + 6^2 = 61$.

2.4.7 $\frac{1}{100}$.

2.4.8 In one hour John does $\frac{1}{2}$ of the job. In one hour Bill does $\frac{1}{3}$. Thus in one hour they together accomplish $\frac{1}{2} + \frac{1}{3} = \frac{5}{6}$ of the job. Thus in $\frac{6}{5}$ of an hour they accomplish the job. Since $\frac{6}{5} \cdot 60 = 72$, it takes them **72** minutes to finish the job.

2.4.9 We have $1\frac{7}{8} = \frac{15}{8}$ and

$$16 \div \frac{15}{5} = \frac{16}{1} \cdot \frac{8}{15} = \frac{128}{15} = 8\frac{8}{15},$$

so she is able to wrap **8** gifts.

2.4.10 Proceeding from the innermost fraction one easily sees that

$$\cfrac{1}{2 - \cfrac{1}{2 - \cfrac{1}{2 - \cfrac{1}{2}}}} = \cfrac{1}{2 - \cfrac{1}{2 - \cfrac{2}{3}}} = \cfrac{1}{2 - \cfrac{3}{4}} = \frac{4}{5}.$$

2.4.11 $\frac{43}{30}$

2.4.12 Of the **100** students, only one is male. He is **2%** of the on-campus population. Thus the whole on-campus population consists of **50** students, so there are **100 − 50 = 50** off-campus students

2.4.13 Observe that $5\frac{1}{2} = \frac{11}{2} = \frac{22}{4}$. Therefore $\frac{21}{4}$ miles are at the rate of **40ᶜ**. The trip costs **\$0.85 +** **\$.40 · 21 = \$.85 + \$8.40 = \$9.25**.

2.5.1

1. +8

2. −26

3. −26

4. 0

5. −303

6. 0

7. +4

8. −β100

2.5.2 Here is one possible answer.

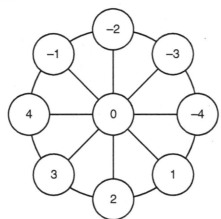

2.5.3 Here is one possible answer.

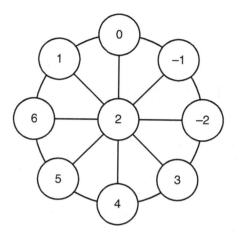

2.5.4 Here are two answers.

−1	×	+1	=	−1
×	■	×	■	×
−1	×	−1	=	+1
=	■	=	■	=
+1	×	−1	=	−1

−1	×	−1	=	+1
×	■	×	■	×
+1	×	−1	=	−1
=	■	=	■	=
−1	×	+1	=	−1

2.5.5 We have,

$$\frac{a^3 + b^3 + c^3 - 3abc}{a^2 + b^2 + c^2 - ab - bc - ca} = \frac{(2)^3 + (-3)^3 + (5)^3 - 3(2)(-3)(5)}{(2)^2 + (-3)^2 + (5)^2 - (2)(-3) - (-3)(5) - (5)(2)}$$

$$= \frac{8 - 7 + 125 + 90}{4 + 9 + 25 + 6 + 15 - 10}$$

$$= \frac{196}{49}$$

$$= 4.$$

2.6.1 We have,

$$\frac{x}{y+z} + \frac{y}{z+x} + \frac{z}{x+y} - \left(\frac{x}{y} + \frac{y}{z} + \frac{z}{x}\right)^2$$

$$= \frac{-1}{2+(-3)} = \frac{2}{-3+(-1)} + \frac{-3}{-1+2} + \left(\frac{-1}{2} + \frac{2}{-3} + \frac{-3}{-1}\right)^2$$

$$= 1 - \frac{1}{2} - 3 + \left(-\frac{1}{2} - \frac{2}{3} + 3\right)^2$$

$$= -2 - \frac{1}{2} + \left(-\frac{1}{2} - \frac{2}{3} + 3\right)^2$$

$$= -\frac{4}{2} - \frac{1}{2} + \left(-\frac{3}{6} - \frac{4}{6} + \frac{18}{6}\right)^2$$

$$= -\frac{5}{2} - \left(\frac{11}{6}\right)^2$$

$$= -\frac{5}{2} - \frac{121}{36}$$

$$= -\frac{90}{36} - \frac{121}{36}$$

$$= -\frac{211}{36}$$

2.6.2

1. $2x + 3y = 2\left(-\frac{2}{3}\right) + 3\left(\frac{3}{5}\right) = -\frac{4}{3} + \frac{9}{5} = -\frac{20}{15} + \frac{27}{15} = \frac{7}{15}$.

2. $xy - x - y = \left(-\frac{2}{3} \cdot \frac{3}{5}\right) - \left(\frac{2}{3}\right) - \frac{3}{5} = -\frac{2}{5} + \frac{2}{3} - \frac{3}{5} = -\frac{6}{15} + \frac{10}{15} - \frac{9}{15} = -\frac{5}{15} = -\frac{1}{3}$.

3. $x^2 + y^2 = \left(-\frac{2}{3}\right)^2 + \left(\frac{3}{5}\right)^2 = \left(-\frac{2}{3}\right)\left(-\frac{2}{3}\right) + \left(\frac{3}{5}\right)\left(\frac{3}{5}\right) = \frac{4}{9} + \frac{9}{25} = \frac{100}{225} + \frac{81}{225} = \frac{181}{225}$.

2.6.3 Here are some possible answers. In some cases, there might be more than one answer, as for example, $1 = \frac{44}{44} = \frac{4+4}{4+4} = \frac{4^4}{4^4}$, etc.

$$1 = \frac{44}{44}$$

$$3 = \frac{4+4+4}{4}$$

$$5 = \frac{4 \cdot 4 + 4}{4}$$

$$7 = \frac{44}{4} - 4$$

$$2 = \frac{4 \cdot 4}{4+4}$$

$$4 = 4 + 4 \cdot (4-4)$$

$$6 = 4 + \frac{4+4}{4}$$

$$8 = 4 \cdot 4 - 4 - 4$$

$$9 = \frac{4}{.4} - \frac{4}{4} \qquad\qquad 10 = \frac{44-4}{4}$$

$$11 = \frac{4}{.4} + \frac{4}{4} \qquad\qquad 12 = \frac{44+4}{4}$$

$$13 = 4! - \frac{44}{4} \qquad\qquad 14 = 4 \cdot (4 - .4) - .4$$

$$15 = \frac{44}{4} + 4 \qquad\qquad 16 = 4 \cdot 4 + 4 - 4$$

$$17 = 4 \cdot 4 + \frac{4}{4} \qquad\qquad 18 = \frac{4}{.4} + 4 + 4$$

$$19 = 4! - 4 - \frac{4}{4} \qquad\qquad 20 = \frac{4}{.4} + \frac{4}{.4}$$

2.6.4 To do this problem correctly, you must compare all combinations of numbers formed with three fours.

Observe that

$$(4^4)^4 = 256^4 = 4294967296,$$

$$44^4 = 3748096, \quad 4^{44} = 309485009821345068724781056,$$

and that

$$4(4^4) = 4^{256}$$

$$= 13407807929942597099574024998205846127479365820592393377723561443721764030073 5 \rightarrow$$

$$\rightarrow 4697680187429816690342769003185818648605085375388281194656994643364900608409 6$$

where we have broken the number so that it fits the paper.

Note: This last calculation was performed with a computer program. How does one actually check that the computer is right? At any rate, since each time we multiply a given quantity by **4** the quantity increases, we must have $4^{44} < 4^{256} = 4^{4^4}$.

2.6.5 $\frac{5}{36}$

2.6.6 $0.1111\ldots = 0.\overline{1} = \frac{1}{3}(0.333333\ldots) = \frac{1}{3} \cdot \frac{1}{3} = \frac{1}{9}$.

2.6.7 From the text, we know that $0.\overline{09} = \frac{1}{11}$, and therefore $121(0.\overline{0.9}) = 121 \cdot \frac{1}{11} = 11$.

2.6.8 $\sqrt{2} + \sqrt{3} + \sqrt{5} \approx 5.38$

2.6.9 $\sqrt{2} \cdot \sqrt{3} \cdot \sqrt{5} \approx 5.48$

2.6.10 No, this is not always true. For example, $\sqrt{1} + \sqrt{3} = 1 + \sqrt{3} \approx 2.73$, but $\sqrt{1+3} = \sqrt{4} = 2$. But, is it *ever* true? For $a = 0$, $\sqrt{a} + \sqrt{b} = \sqrt{0} + \sqrt{b} = \sqrt{b}$ and $\sqrt{a+b} = \sqrt{0+b} = \sqrt{b}$, so it is true in this occasion.

CHAPTER 3 ANSWERS

3.1.1 After buying the marbles, he has $+a$ marbles. After winning b marbles, he now has $a + b$ marbles. After losing c marbles, he ends up with $a + b - c$. In the end, he has $a + b - c$ marbles.

3.1.2

1. 0

2. $-6a$

3. $-10a$

4. $6a$

5. $-4a + 4b$

3.1.3 $2a - 2b + 2c - 2d + 2a^2 - 2c^2$.

3.1.4

1. $4a + 4b + 4c$

2. $4a + 2c$

3. $5a + 5b + 5c$

4. $a - b + c$

5. $2x^2 + 2$

6. $x^3 + x^2 + 1$

7. $6x^3 - 3x^2 - 3x - 2$

8. $-\dfrac{9a}{8}$

3.1.5 Since $(2x^2 + x - 2) + (-x^2 - x + 2) = x^2$, the value is $2^2 = 4$.

3.1.6 Yes, this is always true.

3.1.7 No, this is not always true. For example, if $a = 10$, then $a + a = 10 + 10 = 20$, but $2a^2 = 2(10)^2 = 2 \cdot 100 = 200$. Observe, however, that $0 + 0 = 2 \cdot 0^2$ and that $1 + 1 = 2 \cdot 1^2$, so the assertion is true in these two cases. That it is *only* true in these two cases is more difficult to prove, and requires use of quadratic equations.

3.1.8 No, this is not always true. For example, if $a = 10$, then $a + a^2 = 10 + 100 = 110$, but $2a^3 = 2(10)^3 = 2 \cdot 1000 = 2000$. Observe, however, that $0 + 0^2 = 2 \cdot 0^3$, that $1 + 1^2 = 2 \cdot 1^3$, and that $\left(-\frac{1}{2}\right) + \left(-\frac{1}{2}\right)^2 = 2 \cdot \left(-\frac{1}{2}\right)^3$ so the assertion is true in these three cases. That it is *only* true in these three cases is more difficult to prove, and requires use of quadratic equations.

3.1.9 Letting $x = 1$, one gathers that

$$243 = 3^5 = (8(1) - 5)^5 = A + B + C + D + E + F.$$

3.1.10 Here are a few. Many more are possible.

1. $3xy - x^2 y$

2. $-x^2 y + 3yx$

3. $3yx - yx^2$

4. $3xy - yx^2$

5. $-yx^2 + 3xy$

3.2.1 Since $b - c$ inches are cut from $a + b$, we must subtract $b - c$ from $a + b$, giving

$$(a + b) - (b - c) = a + b - b + c = a + c$$

therefore $a + c$ inches remain.

3.2.2 Since there are x people to share the cost of D, the share of each person is $\frac{D}{x}$. If p persons disappear, there remain $x - p$ persons, and the share of each would increase to $\frac{D}{x-p}$. This means they will have to pay

$$\frac{D}{x-p} - \frac{D}{x}$$

more.

3.2.3 $2b + 2c + 2c^2 + 2d^2$.

3.2.4 $-a + 5b - c$.

3.2.5

1. $-2a + 2c$

2. $3a + 3b + 3c$

3. $a + 3b + c$

4. $x3 - 3x2 + 2x - 3$

5. $-2x^3 - x^2 + 3x - 2$

6. $9.3a - 3.1b$

7. $\frac{11b - 4}{a}$

8. $10 - \frac{8}{a}$

9. $10\clubsuit$

3.2.6 We have

$$\begin{aligned}
\frac{a}{3} - \frac{a^2}{5} + \frac{a}{2} + \frac{5a^2}{6} &= \frac{a}{3} + \frac{a}{2} - \frac{a^2}{5} + \frac{5a^2}{6} \\
&= \frac{2a}{6} + \frac{3a}{6} + \frac{6a^2}{30} + \frac{25a^2}{30} \\
&= \frac{5a}{6} - \frac{19a^2}{30} \\
&= \frac{25a}{30} + \frac{19a^2}{30} \\
&= \frac{25a + 19a^2}{30}
\end{aligned}$$

3.2.7

1. $-2x - 2y$

2. $3x - 5$

3. $4x2 + 2x - 5$

3.2.8 $x - y$ by commutativity is $-y + x$, and so 1 matches with B. By the subtraction rule, $x - (-y) = x + (+y) = x + y$, and so 2 matches with C. Finally, using the distributive law and the commutative law, $-(x - y) = -x + y = y - x$, and so 3 matches with A.

3.2.9 Mary starts with A, and her uncle Bob gives her A more so that she now has $A + A = 2A$. Her aunt Rita gives her 10 more, so that now she has $2A + 10$. She pays B dollars in fines, therefore she is left with $2A + 10 - B$. She spent 12 in fuel, so now she has the following total:

$$2A + 10 - B - 12 = 2A - B - 2,$$

3.2.10

1. $(2 - 4)x - 5(-6)x = -2x + 30x = 28x$

2. $5 - 2(2x) = 5-4x$

3. $(5 - 2)(1 - 2^2)^3 x = (3)(1 - 4)x = (3)(-3)^3 x = (3)(-27)x = -81x$

3.2.11

1. $(6t) + x = 6t + x$, so the parentheses are redundant.

2. In $6(t + x)$, the parentheses are needed.

3. In $(ab)2$, the parentheses are needed.

4. In $(ab - c)2$, the parentheses are needed.

3.2.12

1. In $\frac{a - 2b}{c} = \frac{a}{c} + \boxed{}$, the $\boxed{}$ is $-\frac{2b}{c}$.

2. In $\frac{t - 3}{t - 3} = (t + 3)\boxed{}$, the $\boxed{}$ is $-\frac{1}{t - 3}$.

3.2.13

1. In $1 - t = \boxed{}t + 1$, the $\boxed{}$ is -1 or simply, $-$.

2. In $3x + y - t = \boxed{} + 3x$ the $\boxed{}$ is $y - t$.

3.2.14 The opposite sought is

$$-(4 - y + 2) = -4 + y - 2.$$

3.2.15 The additive inverse sought is

$$-\left(\frac{-a}{-(b - c)}\right) = -\left(\frac{a}{(b - c)}\right) = -\frac{a}{b - c}.$$

CHAPTER 4 ANSWERS

4.1.1 $-8x\ yz$.

4.1.2 $24abc$.

4.1.3 $\frac{4096}{625}$.

4.1.4 $\frac{4}{625}$.

4.1.5 The product equals

$$\frac{4(4^5)}{3(3^5)} \cdot \frac{6(6^5)}{2(2^5)} = 4(4^5) = 2^{12},$$

so $n = 12$.

4.1.6 We have

$$3^{2001} + 3^{2002} + 3^{2003} = 3^{2001}(1 + 3 + 3^2) = (13)3^{2001},$$

therefore $a = 13$.

4.1.7 $(a^x b^y)\left(\dfrac{b^{2x}}{a^{-y}}\right) = a^{x-(-y)}b^{y+2x} = a^{x+y}b^{y+2x}$.

4.1.8 x.

4.1.9 $\dfrac{1}{x^3}$.

4.1.10 $s - 1$.

4.1.11 $\dfrac{1}{(s-1)^3}$.

4.1.12 $\dfrac{a^4}{x^2}$.

4.1.13 1.

4.1.14 We have,

$$\frac{6^5+6^5+6^5+6^5+6^5+6^5+6^5+6^5+6^5+6^5}{3^3+3^3+3^3+3^3+3^3} = \frac{10 \cdot 6^5}{5 \cdot 3^3}$$
$$= \frac{2 \cdot 6^5}{3^3}$$
$$= \frac{2 \cdot 2^5 \cdot 3^5}{3^3}$$
$$= 2 \cdot 2^5 \cdot 3^2$$
$$= 2 \cdot 32 \cdot 9$$
$$= 576.$$

4.1.15 We have,

$$\frac{a^8}{b^9} \div \frac{(a^2b)^3}{b^{20}} = \frac{a^8}{b^9} \cdot \frac{b^{20}}{(a^2b)^3}$$

$$= \frac{a^8}{b^9} \cdot \frac{b^{20}}{a^6b^3}$$

$$= \frac{a^8 b^{20}}{a^6 b^{12}}$$

$$= a^{8-6} b^{20-12}$$

$$= a^2 b^8.$$

So $m = 2$ and $n = 11$.

4.2.1

1. $\frac{1}{x^9}$

2. x^9

3. $\frac{1}{x}$

4. $\frac{y^6 z^9}{x^3}$

5. $\frac{a^9}{b^6}$

6. $a^2 b^5$

4.2.2 We have,

$$(a^{x+3}b^{y-4})(a^{2x-3}b^{-2y+5}) = a^{x+3+\,2x-3}b^{y-4-2y+5} = a^{3x}b^{1-y}.$$

4.2 We have

$$\frac{a^x b^y}{a^{2x} b^{-2y}} = a^{x-2x} b^{y-(-2y)} = a^{-x} b^{3y}.$$

4.2.4 $\frac{x^8}{y^{12}}$.

4.2.5 $2^8 = 256$.

4.2.6 4.

4.2.7 $\frac{(2^4)^8}{(4^8)^2} = \frac{2^{32}}{((2^2)^8)^2} = \frac{2^{32}}{2^{32}} = 1.$

4.3.1

1. $x^3 + x^2 + x$

2. $2x^3 - 4x^2 + 6x$

3. $4x^2 + 4x + 1$

4. $4x^2 - 1$

5. $4x^2 - 4x + 1$

4.3.2 $x^2 + 1 - 7x^3 + 4x - 3x^5.$

4.3.3 $4x^5 - 8x^4 + 8x^3 + 2x^2$.

4.3.4

1. $-9x^4 + 3x^3 - 2x^2 - 2x$

2. $-2x^3 - 4x^2 - 2x$

3. $a^3cb - 2ac^2b$

4. $x^2 - 4y^2 + 3zx - 6zy$

4.3.5 We have,

$$(2x + 2y + 1)(x - y + 1) = (2x + 2y + 1)(x) + (2x + 2y + 1)(-y) + (2x + 2y + 1)(1)$$
$$= 2x^2 + 2xy + x - 2xy - 2y^2 - y + 2x + 2y + 1$$
$$= 2x^2 + 3x - 2y^2 + y + 1.$$

4.3.6 $10x$.

4.3.7 -2.

4.3.8 $x^2 + zx - y^2 + 3zy - 2z^2$.

4.3.9 We have,

$$(a + b + c)(a^2 + b^2 + c^2 - ab - bc - ca) = a^3 + ab^2 + ac^2 - a^2b - abc - ca^2 + ba^2 + b^3 + bc^2 - ab^2$$
$$- b^2c - bca + a^2c + b^2c + c^3 - abc - bc^2 - c^2a$$
$$= a^3 + b^3 + c^3 - 3abc.$$

4.3.10 From the distributive law we deduce that

$$(a + b + c)(x + y + z) = ax + ay + az + bx + by + bz + cx + cy + cz.$$

4.3.11 Let $2a$ be one of the integers and let $2b$ be the other integer. Then $(2a)(2b) = 2(2ab)$, which is twice the integer $2ab$ and therefore it is even.

4.3.12 Let $2a + 1$ be one of the integers and let $2b + 1$ be the other integer. Then

$$(2a + 1)(2b + 1) = 4ab + 2a + 2b + 1$$
$$= 2(2ab) + 2a + 2b + 1$$
$$= 2(2ab + a + b) + 1$$
$$= 2m + 1$$

where $m = 2ab + a + b$. Since m is an integer, the equality $(2a + 1)(2b + 1) = 2m + 1$ shows that the product $(2a + 1)(2b + 1)$ leaves remainder 1 upon division by 2, that is, the product is odd.

4.3.13 Let $4a + 3$ be one of the integers and let $4b + 3$ be the other integer. Then

$$(4a + 3)(4b + 3) = 16ab + 12a + 12b + 9$$
$$= 16ab + 12a + 12b + 8 + 1$$
$$= 4(4ab) + 4(3a) + 4(3b) + 4(2) + 1$$
$$= 4(4ab + 3a + 3b + 2) + 1$$
$$= 4m + 1,$$

where $m = 4ab + 3a + 3b + 1$. Since m is an integer, the equality $(4a + 3)(4b + 3) = 4m + 1$ shows that the product $(4a + 3)(4b + 3)$ leaves remainder **1** upon division by **4**.

4.3.14 Let $3a + 2$ and $3b + 2$ be two numbers leaving remainder **2** upon division by **3**. Then

$$(3a + 2)(3b + 2) = 9ab + 6a + 6b + 4 = 3(3ab + 2a + 2b + 1) + 1,$$

which is a number leaving remainder **1** upon division by **3**.

4.3.15 Let $5a + 2$ and $5b + 2$ be two numbers leaving remainder **2** upon division by **5**. Then

$$(5a + 2)(5b + 2) = 25ab + 10a + 10b + 4 = 5(5ab + 2a + 2b) + 4,$$

which is a number leaving remainder **4** upon division by **5**.

4.3.16 Let $5a + 3$ and $5b + 3$ be two numbers leaving remainder **3** upon division by **5**. Then

$$(5a + 3)(5b + 3) = 25ab + 15a + 15b + 9 = 5(5ab + 2a + 2b + 1) + 4,$$

which is a number leaving remainder **4** upon division by **5**.

4.3.17 Observe that $x^2 + x = 1$. Therefore $x^4 + x^3 = x^2$ and $x^3 + x^2 = x$. Thus

$$x^4 + 2x^3 + x^2 = x^4 + x^3 + x^3 + x^2 = x^2 + x = 1.$$

4.3.18 We have

$$2x(x + 1) - x^2 - x(x - 1) = 2x^2 + 2x - x^2 - x^2 + x$$

$$= 3x.$$

Putting $x = 11112$, one obtains from the preceding part that

$$2 \cdot (11112) \cdot (11113) - 11112^2 - (11111) \cdot (11112) = 3 \cdot 11112 = 33336.$$

4.4.1 $9 - 6y + y^2$.

4.4.2 $25 - 20x^2 + 4x^4$.

4.4.3 $4a^2b^4 - 12ab^2c^3d^4 + 9c^6d^8$.

4.4.4 $x^2 + 8xy + 4y^2$.

4.4.5 $x^2 + 4y^2$.

4.4.6 We have,

$$(xy - 4y)^2 = (xy)^2 - 2(xy)(4y) + (4y)^2 = x^2y^2 - 8xy^2 + 16y^2.$$

4.4.7 We have,

$$(ax + by)^2 = (ax)^2 + 2(ax)(by) + (by)^2 = a^2x^2 + 2abxy + b^2y^2.$$

4.4.8 $2a^2 + 8$.

4.4.9 $8a$.

4.4.10 $a^2 + 4ab + 6ac + 4b^2 + 12bc + 9c^2$.

4.4.11 $a^2 + 4ab - 2ac + 4b^2 - 4bc + c^2$.

4.4.12 Observe that

$$36 = \left(x + \frac{1}{x}\right)^2 = x^2 + 2 + \frac{1}{x^2} \Rightarrow x^2 + \frac{1}{x^2} = 36 - 2 = 34.$$

4.4.13 We have

$$(10^{2002} + 25^2) - (10^{2002} - 25)^2 = 10^{4004} + 2 \cdot 25 \cdot 10^{2002} + 25^2$$
$$- (10^{4004} - 2 \cdot 25 \cdot 10^{2002} + 25^2)$$
$$= 4 \cdot 25 \cdot 10^{2002}$$
$$= 100 \cdot 10^{2002} = 10^{2004},$$

therefore $n = 2004$.

4.4.14 We have

$$4x^2 + 9y^2 = 4x^2 + 12xy + 9y^2 - 12xy$$
$$= (2x + 3y)^2 - 12xy$$
$$= 3^2 - 12(4)$$
$$= 9 - 48$$
$$= -39.$$

4.4.15 Let the two numbers be x, y. Then $x - y = 3$ and $xy = 4$. Therefore

$$9 = (x - y)^2 = x^2 - 2xy + y^2 = x^2 - 8 + y^2 \Rightarrow x^2 + y^2 = 17.$$

4.4.16 We have

$$16 = (x + y)^2 = x^2 + 2xy + y^2 = x^2 - 6 + y^2 \Rightarrow x^2 + y^2 = 22.$$

4.4.17 From Exercise 4.4.16, $x^2 + y^2 = 22$. Observe that $2x^2 y^2 = 2(xy)^2 = 2(-3)^2 = 18$. Therefore

$$484 = (x^2 + y^2)^2 = x^4 + 2x^2 y^2 + y^4 = x^4 + 18 + y^4 \Rightarrow x^4 + y^4 = 466.$$

4.4.18 We have,

$$(x + y + z)^2 = ((x + y) + z)^2$$
$$= (x + y)^2 + 2(x + y)z + z^2$$
$$= x^2 + 2xy + y^2 + 2xz + 2yz + z^2$$
$$= x^2 + y^2 + z^2 + 2xy + 2yz + 2zx.$$

4.4.19 We have

$$(x + y + z + w)^2 = (x + y)^2 + 2(x + y)(z + w) + (z + w)^2$$
$$= x^2 + 2xy + y^2 + 2xz + 2xw + 2yz + 2yw + z^2 + 2zw + w^2$$
$$= x^2 + y^2 + z^2 + w^2 + 2xy + 2xz + 2xw + 2yz + 2yw + 2zw.$$

4.5.1 We have

$$(a + 4)^2 - (a + 2)^2 = ((a + 4) - (a + 2))((a + 4) + (a + 2))$$
$$= (2)(2a + 6)$$
$$= 4a + 12.$$

We may also use the square of a sum identity:

$$(a+4)^2 - (a+2)^2 = (a^2 + 8a + 16) - (a^2 + 4a + 4)$$
$$= 4a + 12,$$

like before.

4.5.2 We have,

$$(x-y)(x+y)(x^2+y^2) = (x^2 - y^2)(x^2 + y^2)$$
$$= x^4 - y^4.$$

4.5.3 We have

$$(a+1)^4 - (a-1)^4 = ((a+1)^2 - (a-1)^2)((a+1)^2 + (a-1)^2)$$
$$= ((a+1) - (a-1))((a+1) + (a-1))((a+1)^2 + (a-1)^2)$$
$$= (2)(2a)((a^2 + 2a + 1) + (a^2 - 2a + 1))$$
$$= 4a(2a^2 + 2)$$
$$= 8a^3 + 8a.$$

4.5.4 Multiplying by $1 = 2 - 1$ does not alter the product, and so,

$$(2-1)(2+1)(2^2+1)(2^4+1)(2^8+1)(2^{16}+1) = (2^2-1)(2^2+1)(2^4+1)(2^8+1)(2^{16}+1)$$
$$= (2^4-1)(2^4+1)(2^8+1)(2^{16}+1)$$
$$= (2^8-1)(2^8+1)(2^{16}+1)$$
$$= (2^{16}-1)(2^{16}+1)$$
$$= 2^{32} - 1.$$

4.5.5 We have,

$$(1^2 - 2^2) + (3^2 - 4^2) + \cdots + (99^2 - 100^2) = (1-2)(1+2) + (3-4)(3+4) + \cdots + (99-100)(99+100)$$
$$= -1(1+2) - 1(3+4) + \cdots + -1(99+100)$$
$$= -(1 + 2 + 3 + \cdots + 100)$$
$$= -5050,$$

using the result of Example 1.5.

4.5.6 Put $x = 123456789$. Then

$$(123456789)^2 - (123456787)(123456791) = x^2 - (x-2)(x+2) = x^2 - (x^2 - 4) = 4.$$

4.5.7 Using $x^2 - y^2 = (x - y)(x + y)$,

$(666\ 666\ 666)^2 - (333\ 333\ 333)^2 = (666\ 666\ 666 - 333\ 333\ 333)(666\ 666\ 666 + 333\ 333\ 333)$

$$= (333\ 333\ 333)(999\ 999\ 999)$$

$$= (333\ 333\ 333)(10^9 - 1)$$

$$= 333\ 333\ 333\ 000\ 000\ 000 - 333\ 333\ 333$$

$$= 333\ 333\ 332\ 666\ 666\ 667.$$

4.6.1 $125x^3 + 75x^2 + 15x + 1$.

4.6.2 $x^3 + 4x^2 + 5x + 2$.

4.6.3 $x^3 - 4x^2 - x - 5$.

4.6.4 $x^3 - 7x^2 + 16x - 12$.

4.6.5 We have,

$216 = (a - b)^3 = a^3 - 3a^2b + 3ab^2 - b^3 = a^3 - b^3 - 3ab(a - b) = a^3 - b^3 - 3(3)(6) \Rightarrow a^3 - b^3 = 270.$

4.6.6 We have,

$$216 = (a + 2b)^3 = a^3 + 6a^2b + 12ab^2 + 8b^3 = a^3 + 8b^3 + 6ab(a + 2b)$$

$$= a^3 + 8b^3 + 6(3)(6) \Rightarrow a^3 + 8b^3 = 108.$$

4.7.1 $x^3 - 8$.

4.7.2 $x^3 + 512$.

4.7.3 Notice that this is Exercise 4.6.5. We will solve it here in a different way. We have

$$a^3 - b^3 = (a - b)(a^2 + ab + b^2) = 6(a^2 + 3 + b^2),$$

so the problem boils down to finding $a^2 + b^2$. Now,

$$36 = (a - b)^2 = a^2 - 2ab + b^2 = a^2 - 6 + b^2 \Rightarrow a^2 + b^2 = 42,$$

therefore

$$a^3 - b^3 = 6(a^2 + 3 + b^2) = 6(42 + 3) = 270$$

as was obtained in the solution of Exercise 4.6.5.

CHAPTER 5 ANSWERS

5.1.1

$$\frac{q^2 - pq - pqr}{-q} = \frac{q^2}{-q} - \frac{pq}{-q} - \frac{-pqr}{-q}$$
$$= -q + p + pr$$
$$= -q + p + pr.$$

5.1.2 $2x^2\,y + 3xy^2$.

5.1.3 $102x^4\,y^4$.

5.1.4 $3a^2x + 2ax$.

5.1.5 $12a^5x^4$.

5.1.6 $-x + y + z$.

5.1.7 $-5a + 7b^2$.

5.1.8 $35a^2b^3$.

5.1.9 $- x^3\,yz$.

5.2.1 $x - 1$.

5.2.2 $x + 3$.

5.2.3 $x^2 - x + 1$.

5.2.4 $2x + 4$.

5.2.5 $2x^2 - 2x - 2$.

5.3.1 We have

$$-4a^6b^5 + 6a^3b^8 - 12a^5b^3 - 2a^2b^3 = -2a^2b^3(2a^4b^2 - 3ab^5 + 6a^3 + 1).$$

5.3.2 $4a^3b^4 - 10a^4b^3 = 2a^3b^3(2b - 5a)$.

5.3.3 $\dfrac{9}{16}x^2 - \dfrac{3}{4}x = \dfrac{3}{4}x\left(\dfrac{3}{4}x - 1\right)$.

5.3.4 $-x + 2y - 3z = -1(x - 2y + 3z)$.

5.3.5 Let $2a + 1$ and $2b + 1$ be odd integers. Then

$$2a + 1 + 2b + 1 = 2a + 2b + 2 = 2(a + b + 1),$$

that is, twice the integer $a + b + 1$, and therefore even.

5.3.6 $x^2(x - 1)$.

5.3.7 $5a^3b^2c^4(25cb^3a - 9ba^2 + 1 - 60c^4a - 2c)$.

5.3.8 $5x^3(x^2 - 2a^7 - 3a^3)$.

5.3.9 $x^2(3xy + 4y^3 - 6x^4 - 10x^2)$.

5.3.10 $3m^4p^2q(m^2p^2q - 2p^2x^2\,ym + m^3px - 3mx + 1)$.

5.3.11 $19a^2x^2(2x^3 + 3a^2)$.

5.3.12 $(a + c)(a + b)$.

5.3.13 $(a + c)(a - b)$.

5.3.14 $(y - 1)(1 + y^2)$.

5.3.15 $(2x + 3)(x^2 + 1)$.

5.3.16 $(a + b)(xa + c + by)$.

5.4.1 $(a - 5)(a - 6)$.

5.4.2 $(a - 19)^2$.

5.4.3 $(a^2b^2 + 25)(a^2b^2 + 12)$.

5.4.4 $(x - 11)(x - 12)$.

5.4.5 $(x^2 - 12)(x^2 - 17)$.

5.4.6 $(x + 27)(x + 8)$.

5.4.7 $(5x + 2)(x + 3)$.

5.4.8 $(7x - 3)(2x + 5)$.

5.5.1 We have

$$x^2 - 4y^2 = (x + 2y)(x - 2y) = (3)(-1) = -3.$$

5.5.2 $(x - 2)(x + 2)(x^2 + 4)$.

5.5.3 $(a + b - c)(a + b + c)$.

5.5.4 $x(x - 1)(x + 1)$.

5.5.5 Since $n^3 - 8 = (n - 2)(n^2 + 2n + 2)$ and $n - 2 < n^2 + 2n + 4$, we must have $n - 2 = 1 \Rightarrow n = 3$ and $n^3 - 8 = 19$.

5.6.1 $\frac{2x}{x^2 - 1}$.

5.6.2 $\frac{2}{x^2 - 1}$.

5.6.3 $\frac{x^2 + 4x - 4}{x^2 - 4}$.

5.6.4 $\frac{x^2 + 4}{x^2 - 4}$.

5.6.5 $\frac{3x - 1}{x^3 - x}$.

5.6.6 $\frac{3a^2 x + 3ax - x}{2a^2}$.

5.6.7 $\frac{1}{s^2 - 1}$.

5.6.8 $\frac{2x - 3y}{xy}$.

5.6.9 $\frac{2s^2 - 7s - 6}{s^2 - 4}$.

5.6.10 $\frac{x(2a - 1)}{a^2}$.

5.6.11 We have,

$$x + y = 7 \Rightarrow x^2 + 2xy + y^2 = 49 \Rightarrow x^2 + y^2 = 49 - 2xy = 49 - 2(21) = 7.$$

Observe also that $x^2 y^2 = (xy)^2 = 21^2$. Therefore

$$\frac{1}{x^2} + \frac{1}{y^2} = \frac{y^2 + y^2}{x^2 y^2} = \frac{7}{21^2} = \frac{1}{63}.$$

CHAPTER 6 ANSWERS

6.1.1 $z = -8$.

6.1.2 $z = 0$.

6.1.3 $z = -11$.

6.1.4 $z = -\frac{3}{4}$.

6.1.5 $z = -16$.

6.1.6 $z = -33$.

6.1.7 $z = -36$.

6.1.8 $z = 3a + 3$.

6.1.9 $z = 5$.

6.1.10 $z = (b^2 + a)(a + 1) = b^2a + b^2 + a^2 + a$.

6.1.11 $N = A$.

6.1.12 $N = A^3 + A$.

6.1.13 $N = \frac{Y^2}{X^2}$.

6.1.14 $N = 12A + A^2$.

6.1.15 $N = -17X$.

6.1.16 $N = A^2 + 3A + 2$.

6.1.17 $N = B^7$.

6.1.18 $N = \frac{2}{A}$.

6.1.19 $N = A + 12$.

6.2.1 $x = \frac{17}{18}$.

6.2.2 $x = 6$.

6.2.3 $x = \frac{12}{5}$.

6.2.4 $x = 3a$.

6.2.5 $x = b$.

6.2.6 $x = \frac{c - b}{a}$.

6.2.7 $x = \frac{a - 2}{3}$.

6.2.8 $x = 1$.

6.2.9 $x = \frac{a}{b}$.

6.2.10 $x = \frac{ab}{cd}$.

6.2.11 $x = 5$.

6.2.12 $x = -13$.

6.2.13 $x = 2$.

6.2.14 $x = \frac{6}{5}$.

6.2.15 We have,

$$(x - a)b = (b - x)a \Rightarrow bx - ab = ab - ax \Rightarrow bx = -ax \Rightarrow (a + b)x = 0 \Rightarrow x = 0.$$

6.2.17 If $x = 0.123123123...$, then $1000x = 0.123123123...$ **and** giving $1000x - x = 123$, since the tails cancel out. This results in $x = \dfrac{123}{999} = \dfrac{41}{333}$.

6.3.1 If x is the number, then

$$6x + 11 = 65 \Rightarrow 6x = 54 \Rightarrow x = 9,$$

so the number is **9**.

6.3.2 If x is the number, then

$$11x - 18 = 15 \Rightarrow 11x = 33 \Rightarrow x = 3,$$

so the number is **3**.

6.3.3 If x is the number, then

$$12(x + 3) = 84 \Rightarrow 12x + 36 = 84 \Rightarrow 12x = 48 \Rightarrow x = 4,$$

so the number is **4**.

6.3.4 Let

$$x - 5, x - 4, x - 3, x - 2, x - 1, x, x + 1, x + 2, x + 3, x + 4, x + 5,$$

be the eleven consecutive integers. Then

$$x - 5 + x - 4 + x - 3 + x - 2 + x - 1 + x + x + 1 + x + 2 + x + 3 + x + 4 + x + 5$$
$$= 2002 \Rightarrow 11x = 2002 \Rightarrow x = 182.$$

The numbers are

$$177, 178, 179, 180, 181, 182, 183, 184, 185, 186, 187.$$

6.3.5 If the two numbers are x and $x + 3$, then

$$x + x + 3 = 27 \Rightarrow x = 12,$$

so the numbers are **12** and **15**.

6.3.6 If x is one number, the other is **19−x**. If x exceeds twice $19 - x$ by **1**, then

$$x - 2(19 - x) = 1 \Rightarrow 3x - 38 = 1 \Rightarrow 3x = 39 \Rightarrow x = 13.$$

Thus one number is **13** and the other is **6**.

6.3.7 If Paul has x dollars, then Mary has $x + 20$ and Peter has $x - 30$. Therefore

$$x + x + 20 + x - 30 = 380 \Rightarrow 3x - 10 = 380 \Rightarrow 3x = 390 \Rightarrow x = 130$$

Therefore Paul has \$130, Mary has \$150, and Peter has \$100.

6.3.8 The amount of acid does not change. Therefore

$$0.60 = 0.20(100 + x) \Rightarrow x = 200,$$

thus **200** grams should be added.

6.3.9 If Bob's age is x, then Bill's age is $x - 4$ and Jane's is $2x + 3$. Therefore

$$2x + 3 + x - 4 + x = 27 \Rightarrow 4x - 1 = 27 \Rightarrow 4x = 28 \Rightarrow x = 7.$$

Thus Bob is **7**, Bill is **3**, and Jane is **17**.

6.3.10 The sum of the original 6 numbers is $S = 6 \cdot 4 = 24.$ If the 7th number is x, then

$$\frac{24 + x}{7} = 5,$$

therefore $x = 11$.

6.3.11 If Bob currently has b dollars, then Bill has $5b$ dollars. After giving \$20 to Bob, Bill now has $5b - 20$ and Bob now has $b + 20$. We are given that

$$5b - 20 = 4(b + 20) \Rightarrow 5b - 20 = 4b + 80 \Rightarrow b = 100.$$

So currently, Bill has \$500 and Bob has \$100.

6.3.12 If x is the number, then

$$\frac{6x}{7} - \frac{4x}{5} = 2.$$

Multiplying both sides of this equation by **35** we obtain

$$35\left(\frac{6x}{7} - \frac{4x}{5}\right) = 2(35) \Rightarrow 30x - 28x = 70 \Rightarrow 2x = 70 \Rightarrow x = 35.$$

6.3.13 If one number is x, the larger is $x + 8$. Thus

$$x + 8 + 2 = 3x \Rightarrow x + 10 = 3x \Rightarrow 10 = 2x \Rightarrow 5 = x,$$

so the smallest number is **5** and the larger is **13**.

6.3.14 If one of the numbers is x, the other is $x + 10$. Thus

$$x + x + 10 = 2(10) \Rightarrow 2x + 10 = 20 \Rightarrow 2x = 10 \Rightarrow x = 5.$$

The smaller number is **5** and the larger is **15**.

6.3.15 If x is the amount originally bought, then I spent $2\left(\frac{x}{4}\right) = \frac{x}{2}$ dollars. Since I kept $\frac{x}{5}$, I must have sold $\frac{4x}{5}$ of them, making $2\left(\frac{\frac{4x}{5}}{3}\right) = \frac{8x}{15}$ dollars on this sale. My net gain is thus

$$\frac{8x}{15} - \frac{x}{2} = 2.$$

Multiplying both sides by **30**, we have

$$30\left(\frac{8x}{15} - \frac{x}{2}\right) = 2(30) \Rightarrow 16x - 15x = 60.$$

I originally bought sixty avocados.

6.3.16 If x is the number, then

$$\frac{x}{4} + \frac{x}{6} + \frac{x}{8} = 13.$$

Multiplying both sides by **24**, we have

$$24\left(\frac{x}{4}+\frac{x}{6}+\frac{x}{8}\right)=13\,(24)\Rightarrow 6x+4x+3x=312\Rightarrow 13x=312\Rightarrow x=24,$$

therefore the number is **24**.

6.3.17 Let x and $x+1$ be the integers. Then

$$\frac{x+1}{5}-\frac{x}{7}=3\Rightarrow 35\left(\frac{x+1}{5}-\frac{x}{7}\right)=3\,(35)$$
$$\Rightarrow 3(x+1)-5x=105$$
$$\Rightarrow 3-2x=105$$
$$\Rightarrow 2x=102$$
$$\Rightarrow x=51.$$

The integers are **51** and **52**.

6.3.18 Let x be the total amount of oranges bought at three for a dollar. On these I spent $\frac{1}{3}\cdot x\cdot 1=\frac{x}{3}$ dollars. I bought $\frac{5x}{6}$ oranges at four for a dollar, thus spending $\frac{1}{4}\cdot\left(\frac{5x}{6}\right)\cdot 1=\frac{5x}{24}$ dollars. Notice that I have bought a total of $x+\frac{5x}{6}=\frac{11x}{6}$ oranges. If I sell all of them at sixteen for six dollars, I make $6\cdot\frac{11x}{6}\cdot\frac{1}{16}=\frac{11x}{16}$ dollars. Thus

$$\frac{11x}{16}-\left(\frac{x}{3}+\frac{5x}{24}\right)=\frac{7}{2}\Rightarrow x=24.$$

Therefore I bought $\left(\frac{11}{6}\right)24=44$ oranges.

6.3.19 Let x be the original price. After a year, the new price will $x-\frac{x}{5}=\frac{4x}{5}$. After another year, the new price will be $\frac{4x}{5}-\frac{1}{6}\cdot\frac{4x}{5}=\frac{2x}{3}$. Therefore we must have

$$\frac{2x}{3}=56000\Rightarrow x=56000\cdot\frac{3}{2}=84000.$$

Thus the original price was $84,000.

6.3.20 Let the integers be n and $n+1$. Then

$$(n+1)^2-n^2=665\Rightarrow 2n+1=665\Rightarrow n=332.$$

Thus the integers are **332** and **333**.

6.3.21 There are ten different ways. We want the number of solutions of

$$5x+10y+25z=50,$$

that is, of

$$x+2y+5z=10,$$

with integer $0\le x\le 10, 0\le y\le 5, 0\le z\le 2$. The table below exhausts all ten possibilities.

z	y	x
2	0	0
1	2	1
1	1	3
1	0	5
0	5	0
0	4	2
0	3	4
0	2	6
0	1	8
0	0	10

6.3.22 Let there be a coins in the purse. There are five stages. The fifth stage is when all the burglars have the same amount of money. Let

$$a_k, b_k, c_k, d_k, e_k$$

be the amount of money that each burglar has, decreasing lexicographically, with the a's denoting the amount of the meanest burglar and e_k denoting the amount of the meekest burglar. Observe that for all k we have

$$a_k + b_k + c_k + d_k + e_k = a$$

On stage five we are given that

$$a_5 = b_5 = c_5 = d_5 = e_5 = \frac{a}{5}.$$

On stage four

$$a_4 = b_4 = c_4 = \frac{a}{5}, \quad d_4 = \frac{a}{5} + \frac{a}{10} = \frac{3a}{10} \quad e_4 = \frac{a}{10}.$$

On stage three we have

$$a_3 = b_3 = \frac{a}{5}, \quad c_3 = \frac{a}{5} + \frac{3a}{20} + \frac{a}{20} = \frac{2a}{5} \quad d_3 = \frac{3a}{20} \quad e_3 = \frac{a}{20}.$$

On stage two we have

$$a_2 = \frac{a}{5}, \quad b_2 = \frac{a}{5} + \frac{a}{5} + \frac{3a}{40} + \frac{a}{40} = \frac{a}{2}, \quad c_2 = \frac{a}{5} \quad d_2 = \frac{3a}{40} \quad e_2 = \frac{a}{40}.$$

On stage one we have

$$a_1 = \frac{a}{5} + \frac{a}{4} + \frac{a}{10} + \frac{3a}{80} + \frac{a}{80} = \frac{3a}{5}, \quad b_1 = \frac{a}{4}, \quad c_1 = \frac{a}{10} \quad d_1 = \frac{3a}{80} \quad e_1 = \frac{a}{80}.$$

Since $a_1 = 240$, we deduce $\frac{3a}{5} = 240 \Rightarrow a = 400.$

Check: On the first stage the distribution is (from meanest to meekest):

240, 100, 40, 15, 5.

On the second stage we have

80, 200, 80 30, 10.

On the third stage we have

80, 80, 160, 60, 20.

On the fourth stage we have

80, 80, 80, 120, 40.

On the fifth stage we have

80, 80, 80, 80, 80.

CHAPTER 7 ANSWERS

7.1.1 $x = 2$ or $x = -2$.

7.1.2 $x = 3$ or $x = -2$.

7.1.3 $x = -3$ or $x = 2$.

7.1.4 $x = 5$ or $x = -1$.

7.1.5 $x = 1$ or $x = -1$.

7.1.6 If eggs had cost x cents less per dozen, I would have saved $\frac{x}{12}$ cents per egg. If eggs had cost x cents more per dozen, I would have lost $\frac{x}{12}$ cents per egg. The difference between these two prices is $2\left(\frac{x}{12}\right)$ per egg. Thus if I buy $x + 3$ eggs and the total difference is 3 cents, I can write

$$(x+3)2\left(\frac{x}{12}\right) = 3,$$

which is to say $x^2 - 3x - 18 = (x - 3)(x + 6) = 0$. Since x has to be a positive number, $x = 3$.

7.1.7 Let x be the original number of people renting the bus. Each person must originally pay $\frac{2300}{x}$ dollars. After six people do not show up there are $x - 6$ people, and each must pay $\frac{2300}{x} + 7.50$ dollars. We need

$$(x-6)\left(\frac{2300}{x} + 7.50\right) = 2300.$$

Rearranging,

$$(x - 6)(2300 + 7.5x) = 2300x \Rightarrow 7.5x^2 - 45x - 13800 = 0 \Rightarrow 75x^2 - 450x - 138000 = 0 \Rightarrow 75$$
$$(x + 40)(x - 46) = 0 \Rightarrow x \in \{-40, 46\}.$$

Since x must be positive, there were **46** people on the bus originally.

7.1.8 One has $x = 6 + \frac{1}{x}$ or $x^2 - 6x - 1 = 0$. Later on you will learn how to solve this equation and see that $x = 3 + \sqrt{10}$. There is some deep mathematics going on here. That the infinite expression makes sense depends on the concept of *convergence*, which is addressed in calculus courses.

7.1.9 We have $x^2 = 2$, and since x is positive, $x = \sqrt{2}$. Like Exercise 7.1.8, this exercise is deep and depends on the concept of convergence.

7.1.10 $(x+1)(x+19)$.

7.1.11 $(x+1)(x+13)$.

7.1.12 $(z+7)^2$.

7.1.13 $(x+1)^2$.

7.1.14 $(x+3)^2$.

7.1.15 $(x-5)^2$.

7.1.16 $(x-3)^2$.

7.1.17 $(x-6)^2$.

7.1.18 $z = 0$ and $z = 2$. Remember the Zero Product property states that for all real numbers x and y:if $xy = 0$, then $x = 0$ or $y = 0$. Therefore $(z-2) = 0$ and $z = 0$.

7.1.18 $n = \frac{-4}{3}$ or $n = 2$.

7.1.20 $t = \frac{9}{5}$ and $t = -2$.

7.1.21. $x = -\frac{9}{5}$ and $x = 2$.

7.1.22. $n = -\frac{3}{2}$ and $n = 13$.

7.2.1. $x^2 + 30x + 225$.

Add $\left(\frac{b}{2}\right)^2$ to complete the square, $x^2 + 30x + \left(\frac{b}{2}\right)^2$ and substitute in $b = 30$. $x^2 + 30x + \left(\frac{30}{2}\right)^2 \rightarrow x^2 + 30x + 15^2$. This quadratic equation can be written in a complete square, $(x+15)^2$. The perfect square was complete with the number **225.**

7.2.2. $x^2 - 6x - 36$.

Add $\left(\frac{b}{2}\right)^2$ to complete the square, $x^2 - 6x + \left(\frac{b}{2}\right)^2$ and substitute in $b = -6$. $x^2 - 6x + \left(\frac{-6}{2}\right)^2 \rightarrow x^2 - 6x + (-3)^2$. This quadratic equation can be written in a complete square, $(x-3)^2$. The perfect square was complete with the number **9.**

7.2.3. $h^2 + 10h + 25$.

Add $\left(\frac{b}{2}\right)^2$ to complete the square, $h^2 + 10h + \left(\frac{b}{2}\right)^2$ and substitute in $b = 10$. $h^2 + 10h + \left(\frac{10}{2}\right)^2 \rightarrow h^2 + 10h + (5)^2$. This quadratic equation can be written in a complete square, $(h+5)^2$. The perfect square was complete with the number **25.**

7.2.4. $x^2 - 16x + 64$.

Add $\left(\frac{b}{2}\right)^2$ to complete the square, $x^2 - 16x + \left(\frac{b}{2}\right)^2$ and substitute in $b = -16$. $x^2 - 16x + \left(\frac{-16}{2}\right)^2 \rightarrow x^2 - 16x + (-8)^2$. This quadratic equation can be written in a complete square, $(x-8)^2$.

The perfect square was complete with the number **64.**

7.2.5. $p^2 + 36p + 324.$

Add $\left(\dfrac{b}{2}\right)^2$ to complete the square, $p^2 + 36p + \left(\dfrac{b}{2}\right)^2$ and substitute in $b = 36.$

$p^2 + 36p + \left(\dfrac{36}{2}\right)^2 \to p^2 + 36p + (18)^2.$ This quadratic equation can be written in a complete square, $(x + 18)^2.$ The perfect square was complete with the number $324.$

7.2.6. $x^2 + 28x + 196.$

Add $\left(\dfrac{b}{2}\right)^2$ to complete the square, $x^2 + 28x + \left(\dfrac{b}{2}\right)^2$ and substitute in $b = 28.$ $x^2 + 28x + \left(\dfrac{28}{2}\right)^2 \to x^2 + 28x + (14)^2.$ This quadratic equation can be written in a complete square, $(x + 14)^2.$ The perfect square was complete with the number $196.$

7.2.7. $x^2 - 26x + 169.$

Add $\left(\dfrac{b}{2}\right)^2$ to complete the square, $x^2 - 26x + \left(\dfrac{b}{2}\right)^2$ and substitute in $b = -26.$

$x^2 - 26x + \left(\dfrac{-26}{2}\right)^2 \to x^2 - 26x + (-13)^2.$ This quadratic equation can be written in a complete square, $(x - 13)^2.$ The perfect square was complete with the number $169.$

7.2.8. $h^2 - 10h + 25.$

Add $\left(\dfrac{b}{2}\right)^2$ to complete the square, $h^2 - 10h + \left(\dfrac{b}{2}\right)^2$ and substitute in $b = -10.$

$h^2 - 10h + \left(\dfrac{-10}{2}\right)^2 \to h^2 - 10h + (-5)^2.$ This quadratic equation can be written in a complete square, $(h - 5)^2.$ The perfect square was complete with the number $25.$

7.2.9. $x^2 + 2x + 1.$

Add $\left(\dfrac{b}{2}\right)^2$ to complete the square, $x^2 + 2x + \left(\dfrac{b}{2}\right)^2$ and substitute in $b = 2.$ $x^2 + 2x + \left(\dfrac{2}{2}\right)^2 \to x^2 + 2x + (1)^2.$

This quadratic equation can be written in a complete square, $(x + 1)^2.$ The perfect square was complete with the number $1.$

7.2.10. $x^2 - 2x + 1.$

Add $\left(\dfrac{b}{2}\right)^2$ to complete the square, $x^2 - 2x + \left(\dfrac{b}{2}\right)^2$ and substitute in $b = -2.$ $x^2 - 2x + \left(\dfrac{-2}{2}\right)^2 \to x^2 - 2x + (-1)^2.$

This quadratic equation can be written in a complete square, $(x - 1)^2.$ The perfect square was complete with the number $1.$

7.3.1 $p \approx 1.11$ or $p = -8.11.$

Use the quadratic formulat to solve $p^2 + 7p - 9 = 0$ *and* substitute in $a = 1, b = 7,$ and $c = -9.$

$$p = \frac{-b \pm \sqrt{b^2 - 4ac}}{2a} = \frac{-7 \pm \sqrt{7^2 - 4(1)(-9)}}{2(1)} = \frac{-7 \pm \sqrt{49 + 36}}{2} = \frac{-7 \pm \sqrt{85}}{2}.$$

7.3.2 $x \approx 1.84$ *or* $x \approx -1.09.$

Use the quadratic formulat to solve $4x^2 - 3x - 8 = 0$ *and* substitute in $a = 4, b = -3,$ and $c = -8.$

$$x = \frac{-b \pm \sqrt{b^2 - 4ac}}{2a} = \frac{-(-3) \pm \sqrt{(-3)^2 - 4(4)(-8)}}{2(4)} = \frac{3 \pm \sqrt{9 + 128}}{8} = \frac{3 \pm \sqrt{137}}{8}.$$

7.3.3 $f \approx 3.09$ or $f = -0.76$.

Use the quadratic formulat to solve $3f^2 - 7f - 7 = 0$ *and* substitute in $a = 3$, $b = -7$, and $c = -7$.

$$f = \frac{-b \pm \sqrt{b^2 - 4ac}}{2a} = \frac{-(-7) \pm \sqrt{(-7)^2 - 4(3)(-7)}}{2(3)} = \frac{7 \pm \sqrt{49 + 84}}{6} = \frac{7 \pm \sqrt{133}}{8}.$$

7.3.4 $z = -1$ or $z = -2\frac{1}{2}$.

Use the quadratic formulat to solve $2z^2 + 7z + 5 = 0$ and substitute in $a = 2$, $b = 7$, and $c = 5$.

$$z = \frac{-b \pm \sqrt{b^2 - 4ac}}{2a} = \frac{-(7) \pm \sqrt{(7)^2 - 4(2)(5)}}{2(2)} = \frac{-7 \pm \sqrt{49 - 40}}{4} = \frac{-7 \pm \sqrt{9}}{4} = \frac{-7 \pm 3}{4}.$$

$$z = \frac{-7 + 3}{4} = -\frac{4}{4} = -1 \text{ or } z = \frac{-7 - 3}{4} = -\frac{10}{4} = -\frac{5}{2} = -2\frac{1}{2}.$$

7.3.5 $q = -1$.

Use the quadratic formulat to solve $2q^2 + 4q + 2 = 0$ and substitute in $a = 2$, $b = 4$, and $c = 2$.

$$q = \frac{-b \pm \sqrt{b^2 - 4ac}}{2a} = \frac{-(4) \pm \sqrt{(4)^2 - 4(2)(2)}}{2(2)} = \frac{-4 \pm \sqrt{16 - 16}}{4} = \frac{-4 \pm 0}{4} = \frac{-4}{4} = -1.$$

7.3.6 $n \approx 4.91$ or $n \approx -0.41$.

Use the quadratic formulat to solve $2n^2 - 9n - 4 = 0$, and substitute in $a = 2$, $b = -9$, and $c = -4$.

$$n = \frac{-b \pm \sqrt{b^2 - 4ac}}{2a} = \frac{-(-9) \pm \sqrt{(-9)^2 - 4(2)(-4)}}{2(2)} = \frac{9 \pm \sqrt{81 + 32}}{4} = \frac{9 \pm \sqrt{113}}{4}.$$

7.3.7 $x \approx 1.92$ or $x \approx -1.17$.

Use the quadratic formulat to solve $4x^2 - 3x - 9 = 0$, and substitute in $a = 4$, $b = -3$, and $c = -9$.

$$x = \frac{-b \pm \sqrt{b^2 - 4ac}}{2a} = \frac{-(-3) \pm \sqrt{(-3)^2 - 4(4)(-9)}}{2(4)} = \frac{3 \pm \sqrt{9 + 144}}{8} = \frac{3 \pm \sqrt{153}}{8}.$$

7.3.8 $u \approx 0.59$ or $x \approx -2.39$.

Use the quadratic formula to solve $5u^2 + 9u - 7 = 0$, and substitute in $a = 5$, $b = 9$, and $c = -7$.

$$u = \frac{-b \pm \sqrt{b^2 - 4ac}}{2a} = \frac{-(9) \pm \sqrt{(9)^2 - 4(5)(-7)}}{2(5)} = \frac{-9 \pm \sqrt{81 + 140}}{10} = \frac{-9 \pm \sqrt{221}}{10}.$$

7.3.9 $w = 0$ or $w \approx 3.28$ or $w \approx 1.22$.

Notice that this equation can be rewritten as $w(2w^2 - 9w + 8) = 0$, so one solution is $w = 0$. Let's find the rest of possible solutions by using the quadratic formulat to solve $2w^2 - 9w + 8 = 0$, and substitute in $a = 2$, $b = -9$, and $c = 8$.

$$w = \frac{-b \pm \sqrt{b^2 - 4ac}}{2a} = \frac{-(-9) \pm \sqrt{(-9)^2 - 4(2)(8)}}{2(2)} = \frac{9 \pm \sqrt{81 - 64}}{4} = \frac{9 \pm \sqrt{17}}{4}.$$

7.3.10 $m = 0$ or $w \approx -0.27$ or $w \approx -3.73$.

Notice that $2m^3 + 8m^2 + 2m = 0$, can be rewritten as $m(2m^2 + 8m + 2) = 0$, so one solution is $m = 0$. Let's find the rest of possible solutions by using the quadratic formulat to solve $2m^2 + 8m + 2 = 0$, and substitute in $a = 2$, $b = 8$, and $c = 2$.

$$m = \frac{-b \pm \sqrt{b^2 - 4ac}}{2a} = \frac{-(8) \pm \sqrt{(8)^2 - 4(2)(2)}}{2(2)} = \frac{-8 \pm \sqrt{64 - 16}}{4} = \frac{9 \pm \sqrt{48}}{4}.$$

CHAPTER 8 ANSWERS

8.2.1 $x > -2$.

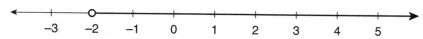

In order to solve this expression, first collect like terms on one side of the expression then manipulate the expression such that the unknown variable x is on one side and the rest of the expression is grouped to the other side of the inequality. The following are the steps to solve this problem.

$$x - \frac{x-6}{4} < 2 + x$$

$$-x + x - \frac{x-6}{4} < 2 + x - x$$

$$4(-\frac{x-6}{4} < 2) \rightarrow -x + 6 < 8$$

$$-x < 2 \rightarrow x > -2.$$

8.2.2 $x \leq -3$.

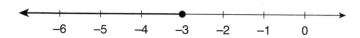

In order to solve this expression, first multiply the right-hand expression. Collect like terms on one side of the expression, then manipulate the expression such that the unknown variable x is on one side and the rest of the expression is grouped to the other side of the inequality. The following are the steps to solve this problem.

$$4(x + 1) \leq x - 5 \rightarrow 4x + 4 \leq x - 5$$

$$-x + 4x + 4 \leq x - 5 - x \rightarrow 3x + 4 \leq -5$$

$$3x + 4 - 4 \leq -5 - 4$$

$$3x \leq -9 \rightarrow \frac{3x \leq -9}{3}$$

$$x \leq -3.$$

8.2.3 $x > 1$.

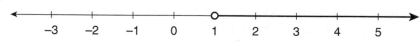

In order to solve this expression, first multiply the right-hand expression. Collect like terms on one side of the expression then manipulate the expression such that the unknown variable x is on one side and the rest of the expression is grouped to the other side of the inequality. The following are the steps to solve this problem.

$$\frac{x-1}{3} - \frac{1-x}{2} > 0 \rightarrow 6(\frac{x-1}{3} - \frac{1-x}{2} > 0)$$
$$2x - 2 - 3 + 3x > 0 \rightarrow 5x - 5 > 0$$
$$5x > 5 \rightarrow x > 1.$$

8.3.1 (A) $x < 75$ Let x be the unknown number, therefore the expression can be written as:

$$2x - 50 < 100 \rightarrow 2x - 50 + 50 < 100 + 50$$
$$2x < 150 \rightarrow \frac{2x < 150}{2} \rightarrow x < 75.$$

8.3.2 $\$10.75 x \leq \$860$ Let x equal the number of hours Dalya works. If she works the maximum number of hours, or **40**, per pay will be **40($\$10.75$)**, which equals **$\$430$** per week. Therefore, the inequality expression is set up so that the wage times the number of hours is equal to or less than **2($\$430$),** for two weeks (biweekly).

8.3.3 $s < 18$ Set up the perimeter of a square formula $p = 4s$ with an inequality or **$72 > 4s$** or **$18 > s$**, by dividing both sides by **4**, which can be rewritten as **$s < 18$**.

8.3.4 The velocity of the ball will be between 30 and 60 feet per second between 0.5 and 1.5 seconds after initial launch. Set up the compound inequality, and then solve for t:

$$30 < 75 - 30t < 60$$
$$30 - 75 < 75 - 75 - 30t < 60 - 75$$
$$-45 < -30t < -15$$
$$\frac{-45}{-30} > \frac{-30t}{-30} > \frac{-15}{-30}$$
$$\frac{3}{2} > t > \frac{1}{2}$$
$$1.5 > t > 0.5$$

or

$$0.5 < t < 1.5.$$

Remember that since we divided by a negative number, we flipped the inequality signs. Also, we rearrange the answer such that the time range is listed between **0.5** and **1.5** seconds.

8.3.5 Robert needs to invest $30,000 at 11% to obtain a $2,700 interest income from his mutual funds account. Set up the simple interest formula equation as $I = Prt$, where I is the interest, P is the initial investment or principal, and r is the rate of interest expressed in decimal, and t is the time in years. Let x be the amount to invest in mutual funds. Then, we will have **$10,000 - x$** left to invest in the simple interest account. The interest on the mutual funds investment will be $x(0.11)(1) = 0.11x$. The interest on the simple interest account will be $(10000 - x)(0.03)(1) = 300 - 0.03x$. Then the total interest is $0.11x + (300 - 0.03x) = 0.08x + 300$. The minimum amount to invest in mutual funds is **$2,700**, so $0.08x + 300 \geq 2700 \rightarrow 0.08x \geq 2400 \rightarrow x \geq 30000$. In other words, Robert will need to invest at least **$30,000** in mutual funds in order to get **$2,700** in interest income.

8.4.1 $x > -5$ or $x < -15$.

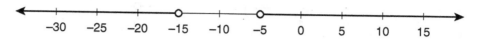

The following are the steps to solve the compound inequalities:

$$x - 10 > -15 \text{ or } x + 1 < -14$$

$$x - 10 + 10 > -15 + 10 \text{ or } x + 1 - 1 < -14 - 1$$

$$x > -5 \text{ or } x < -15.$$

8.4.2 $x > 8$ or $x < -6$.

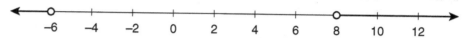

The following are the steps to solve the compound inequalities:

$$x - 15 > -7 \text{ or } x + 19 < 13$$

$$x - 15 + 15 > -7 + 15 \text{ or } x + 19 - 19 < 13 - 19$$

$$x > 8 \text{ or } x < -6.$$

8.4.3 $3 \le x \le 6$.

The following are the steps to solve the compound inequalities:

$$16 \ge x + 10 \ge 13$$

$$16 - 10 \ge x + 10 - 10 \ge 13 - 10$$

$$6 \ge x \ge 3 \rightarrow 3 \le x \le 6.$$

8.4.4 $-6 \ge x \ge -14$.

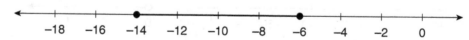

The following are the steps to solve the compound inequalities:

$$11 \ge x + 17 \ge 3$$

$$11 - 17 \ge x + 17 - 17 \ge 3 - 17$$

$$-6 \ge x \ge -14.$$

8.4.5 $8 < y < 10$.

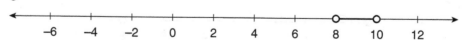

The following are the steps to solve the compound inequalities:

$$-7 > y - 17 > -9$$

$$-7 + 17 > y - 17 + 17 > -9 + 17$$

$$10 > y > 8 \rightarrow 8 < y < 10.$$

8.4.6 $2 < m < 5.$

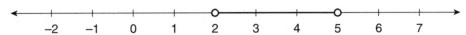

The following are the steps to solve the compound inequalities:

$$-14 > m - 19 > -17$$
$$-14 + 19 > m - 19 + 19 > -17 + 19$$
$$5 > m > 2 \rightarrow 2 < m < 5.$$

8.4.7 $b \leq -1$ or $b > 8.$

The following are the steps to solve the compound inequalities:

$$b - 4 \leq -5 \text{ or } b + 1 > 9$$

$$b - 4 + 4 \leq -5 + 4 \text{ or } b + 1 - 1 > 9 - 1$$

$$b \leq -1 \text{ or } b > 8.$$

8.4.8 $a \geq 12$ or $a < 10.$

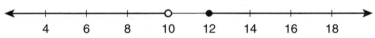

The following are the steps to solve the compound inequalities:

$$a - 5 \geq 7 \text{ or } a < 10$$

$$a - 5 + 5 \geq 7 + 5 \text{ or } a < 10$$

$$a \geq 12 \text{ or } a < 10.$$

8.4.9 $z > 0$ or $z < -10.$

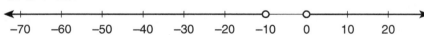

The following are the steps to solve the compound inequalities:

$$z - 8 > -8 \text{ or } z - 12 < -22,$$

$$z - 8 + 8 > -8 + 8 \text{ or } z - 12 + 12 < -22 + 12$$

$$z > 0 \text{ or } z < -10.$$

8.4.10 $-4 < c < 8.$

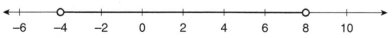

The following are the steps to solve the compound inequalities:

$$-5 > c - 13 > -17$$

$$-5 + 13 > c - 13 + 13 > -17 + 13$$

$$8 > c > -4 \text{ or } -4 < c < 8.$$

8.4.11 $x \geq 16$ or $x \leq 4.$

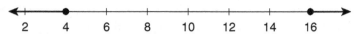

The following are the steps to solve the compound inequalities:

$$x - 1 \geq 15 \text{ or } x + 8 \leq 12$$

$$x - 1 + 1 \geq 15 + 1 \text{ or } x + 8 - 8 \leq 12 - 8$$

$$x \geq 16 \text{ or } x \leq 4.$$

8.4.12 $y \geq 6$ or $y < -6.$

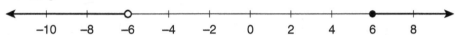

The following are the steps to solve the compound inequalities:

$$y + 9 \geq 15 \text{ or } y + 4 < -2$$

$$y + 9 - 9 \geq 15 - 9 \text{ or } y + 4 - 4 < -2 - 4$$

$$y \geq 6 \text{ or } y < -6.$$

8.4.13 $-3 \leq x \leq 5.$

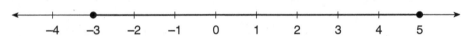

The following are the steps to solve the compound inequalities:

$$-5 \leq 2x + 1 \leq 11$$

$$-5 - 1 \leq 2x + 1 - 1 \leq 11 - 1 \rightarrow -6 \leq 2x \leq 10$$

$$\frac{(-6 \leq 2x \leq 10)}{2} \rightarrow -3 \leq x \leq 5.$$

8.4.14 $-10 \leq n \leq 10.$

The following are the steps to solve the compound inequalities:

$$-3 \leq n + 7 \leq 17 \rightarrow -3 - 7 \leq n + 7 - 7 \leq 17 - 7$$

$$-10 \leq n \leq 10.$$

8.4.15 $-5 < p < 1.$

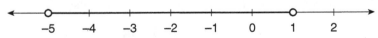

The following are the steps to solve the compound inequalities:

$$17 > p + 16 > 11 \rightarrow 17 - 16 > p + 16 - 16 > 11 - 16$$

$$1 > p > -5 \text{ or } -5 < p < 1.$$

8.4.16 $x < 3$ or $x \geq 4$.

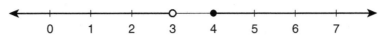

The following are the steps to solve the compound inequalities:

$$x - 2 < 1 \text{ or } x - 12 \geq -8$$

$$x - 2 + 2 < 1 + 2 \text{ or } x - 12 + 12 \geq -8 + 12$$

$$x < 3 \text{ or } x \geq 4.$$

8.4.17 $9 < z \leq 15$.

The following are the steps to solve the compound inequalities:

$$11 < z + 2 \leq 17 \rightarrow 11 - 2 < z + 2 - 2 \leq 17 - 2$$

$$9 < z \leq 15.$$

8.4.18 $-5 > p > -8$.

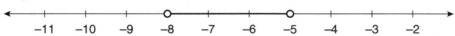

The following are the steps to solve the compound inequalities:

$$-7 > p - 3 > -11 \rightarrow -7 + 3 > p - 3 + 3 > -11 + 3$$

$$-5 > p > -8.$$

8.4.19 $n > 16$ or $v < 14$.

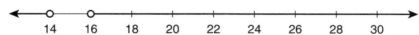

The following are the steps to solve the compound inequalities

$$n - 9 > 7 \text{ or } v - 20 < -6$$

$$n - 9 + 9 > 7 + 9 \text{ or } v - 20 + 20 < -6 + 20$$

$$n > 16 \text{ or } v < 14.$$

8.4.20 $v \leq 1$ or $v > 8$.

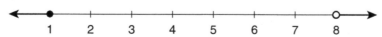

The following are the steps to solve the compound inequalities:

$$v + 6 \leq 7 \text{ or } v + 6 > 14$$

$$v + 6 - 6 \leq 7 - 6 \text{ or } v + 6 - 6 > 14 - 6$$

$$v \leq 1 \text{ or } v > 8.$$

APPENDIX C: ANSWER KEY FOR ODD-NUMBERED REVIEW EXERCISES

1. **(B) $x - 3$.** The excess of a number (x) over another number (3) is another way of saying the difference between two numbers or x and 3.

3. **(B) "The square of a number is reduced by the reciprocal of the number."**

5. **(A) 1** $(-100) \div (10) \div (-10) = (-10) \div (-10) = 1$.

7. **(B)** $\frac{4}{49}$ Remember that division is the same as multiplying a number by its reciprocal:

$$\frac{3}{14} \div \frac{21}{8} = \frac{3}{14} \cdot \frac{8}{21} = \frac{(3)(8)}{(14)(21)} = \frac{(3 \times 1)(2 \times 4)}{(2 \times 7)(3 \times 7)} = \frac{(1)(4)}{(7)(7)} = \frac{4}{49}.$$

9. **(C) 4.** Remember the order of operations or the acrynom **PEMDAS**: **P**arentheses first, **E**xponents, **M**ultiplication and **D**ivision (from left to right), the **A**ddition and **S**ubtraction (also, from left to right). To answer this question, first divide **1024** by **512** then multiply by **2** or $(1024 \div 512) \times 2 = 2 \times 2 = 4$.

11. **(B) 9** $1 - (-2)^3 = 1 - (-8) = 1 + 8 = 9$.

13. **(A)** $t + w$.

15. **(A)** $\frac{13}{42}$. Find the common denominator for all three terms and solve the expression:

$$\frac{1}{2} + \frac{2}{3} - \frac{6}{7} = \frac{(21)(1)}{(21)(2)} + \frac{(14)2}{(14)(3)} - \frac{(6)(6)}{(6)(7)} = \frac{21}{42} + \frac{28}{42} - \frac{36}{42} = \frac{13}{42}.$$

17. **(C) –5.**

19. **(C)** $\frac{9}{10}$. Find the common denominator (12) and solve as follows:

$$\frac{3}{4} + \frac{5}{6} = \frac{(3)(3)}{(3)(4)} + \frac{(2)(5)}{(2)(6)} = \frac{9}{12} + \frac{10}{12} = \frac{19}{12}.$$

21. **(E) none of these.**

23. **(B) $7x - 4$.** First, reduce the terms inside of the parentheses, then collect like terms to reduce the expression: $\left(\frac{20x - 10}{5}\right) - \left(\frac{6 - 9x}{3}\right) = (4x - 2) - (2 - 3x) = (4x + 3x) + (-2 - 2) = 7x - 4$

25. (A) $-6x^2-4x+10$. Multiply through each expression and collect like terms:
$-2(x^2-1)+4(2-x-x^2)=-2x^2+2+8-4x-4x^2=(-2x^2-4x^2)-4x+(2+8)=-6x^2-4x+10$

27. (B) $2a-2b$. $-a+2b+3a-4b=(-a+3a)+(2b-4b)=2a-2b$.

29. (B) $\frac{7a}{12}+\frac{13a^2}{12}$. First, find a common denominator of **12** then solve the expression:

$$\frac{a}{3}+\frac{a^2}{3}+\frac{a}{4}+\frac{3a^2}{4}=\frac{4a}{12}+\frac{4a^2}{12}+\frac{3a}{12}+\frac{9a^2}{12}=\left(\frac{4a}{12}+\frac{3a}{12}\right)+\left(\frac{4a^2}{12}+\frac{9a^2}{12}\right)=\frac{7a}{12}+\frac{13a^2}{12}.$$

31. (B) $2x-3$. $\frac{3x-6}{3}+x-1=\frac{3x}{3}-\frac{6}{3}+x-1=x-2+x-1=$

$$(x+x)+(-2-1)=2x-3.$$

33. (A) $-11x+6$. $\left(\frac{-35x+20}{5}\right)+\left(\frac{12-24x}{6}\right)=(-7x+4)+(2-4x)=$

$$(-7x-4x)+(4+2)=-11x+6.$$

35. (C) $\frac{1}{3}$. $\frac{3^3}{3^4}=3^{(3-4)}=3^{(-1)}=\frac{1}{3}$.

37. (D) 3^7. $3^3\cdot 3^4=3^{(3+4)}=3^7$.

39. (B) 2222. $\frac{1111^5+1111^5+1111^5+1111^5}{1111^4+1111^4}=\frac{4(1111^5)}{2(1111^4)}=$

$$2(1111)^{(5-4)}=2(1111)^1=2(1111)=2222.$$

41. (C) $\frac{b^8}{a}$. $(ab^2)^2\div(a^3b^{-4})=\frac{(a^2b^4)}{(a^3b^{-4})}=a^{(2-3)}b^{(4-(-4))}=a^{-1}b^8=\frac{b^8}{a^1}=\frac{b^8}{a}$.

43. (C) $\frac{799}{25}$. $\frac{1}{2^{-5}}-5^{-2}=2^5-\frac{1}{5^2}=32-\frac{1}{25}=\left(\frac{25}{25}\cdot\frac{32}{1}\right)-\frac{1}{25}=\frac{800}{25}-\frac{1}{25}=\frac{799}{25}$.

45. (A) x^3-8. $(x-2)(x^2+2x+4)=x^3+2x^2+4x-2x^2-4x-8=$
$$x^3+(2x^2-2x^2)+(4x-4x)-8=x^3-8.$$

47. (A) $\frac{9}{64}$. $\frac{2^{-3}}{3^{-2}2^3}=\frac{3^2}{2^{(3+3)}}=\frac{9}{2^6}=\frac{9}{64}$.

49. (D) $x^6y^7z^4$. $(x^4y^3z^2)(x^2y^4z^2)=x^{(4+2)}y^{(3+4)}z^{(2+2)}=x^6y^7z^4$.

51. (B) $4x^2-4x+1$. $(2x-1)(2x+1)-2(2x-1)=(4x^2-2x+2x-1)-4x+2=4x^2-4x+1$.

53. (A) $2a^2+8$. $(a+2)^2+(a-2)^2=(a^2+4a+4)+(a^2-4a+4)=2a^2+8$.

55. (B) x^3-2x^2-7x+2. $(x+2)(x^2-4x+1)=x^3-4x^2+x+2x^2-8x+2=x^3-2x^2-7x+2$.

57. (A) $x(2x-3)$.

59. (D) $(x-1)(x-2)$.

61. **(B)** $x^2 - 2x - 1.$ $(2x^3 - 3x^2 - 4x - 1) \div (2x + 1) = \dfrac{(2x^3 - 3x^2 - 4x - 1)}{(2x + 1)} =$

$$\frac{(2x+1)(x^2 - 2x - 1)}{(2x + 1)} = x^2 - 2x - 1.$$

63. **(D)** $6x^2.$ $\dfrac{(6x^2)(2x)}{2x} = 6x^2.$

65. **(B)** $x + 7.$ $(x^2 + 5x - 14) \div (x - 2) = \dfrac{(x^2 + 5x - 14)}{(x - 2)} = \dfrac{(x - 2)(x + 7)}{(x - 2)} = x + 7.$

67. **(D)** $x^2 - 3x + 9.$ $(x^3 + 27) \div (x + 3) = \dfrac{(x^3 + 27)}{(x + 3)} = \dfrac{(x + 3)(x^2 - 3x + 9)}{(x + 3)} = x^2 - 3x + 9.$

69. **(D)** $x^{12}.$ $\dfrac{(x^9)(x^6)}{x^3} = \dfrac{x^{15}}{x^3} = x^{12}.$

71. **(C)** $(x + 3)(x - 4).$

73. **(C)** $x(x - 3)(x - 9).$ $x^3 - 12x^2 + 27x = x(x^2 - 12x + 27) = x(x - 3)(x - 9).$

75. **(A)** $a(a + 8)^2.$ $a^3 + 16a^2 + 64a = a(a^2 + 16a + 64) = a(a + 8)^2.$

77. **(B)** $2x^2 - x - 3.$ $(6x^3 + x^2 - 11x - 6) \div (3x + 2) = \dfrac{(6x^3 + x^2 - 11x - 6)}{(3x + 2)} =$

$$\frac{(3x+2)(2x^2 - x - 3)}{(3x + 2)} = 2x^2 - x - 3.$$

79. **(B)** $\dfrac{5x}{x^2 + x - 6}.$ $\dfrac{2}{x-2} + \dfrac{3}{x+3} = \dfrac{(x+3)}{(x+3)} \cdot \dfrac{2}{x-2} + \dfrac{(x-2)}{(x-2)} \cdot \dfrac{3}{x+3} =$

$$\frac{2x+6}{(x+3)(x-2)} + \frac{3x-6}{(x+3)(x-2)} = \frac{5x}{(x+3)(x-2)} = \frac{5x}{x^2 + x - 6}.$$

81. **(A)** $\dfrac{a}{a+1}.$ $\dfrac{a^2}{a^2 + a} = \dfrac{a(a)}{a(a+1)} = \dfrac{a}{a+1}.$

$$\frac{2x+4}{x^2 - x - 6} + \frac{3x-9}{x^2 - x - 6} = \frac{5x-5}{x^2 - x - 6}.$$

83. **(B)** $\dfrac{13+x}{x^2 - x - 6}.$ $\dfrac{2}{x-3} - \dfrac{3}{x+2} = \left(\dfrac{x+2}{x+2} \cdot \dfrac{2}{x-3} \right) - \left(\dfrac{x-3}{x-3} \cdot \dfrac{3}{x+2} \right) =$

$$\frac{2x+4}{x^2 - x - 6} - \frac{3x-9}{x^2 - x - 6} = \frac{13+x}{x^2 - x - 6}.$$

85. **(C)** $x = 2b.$ $\dfrac{ax}{b} = 2a = \dfrac{b \cdot ax}{b} = b \cdot 2a \rightarrow ax = 2ab \rightarrow \dfrac{ax}{a} = \dfrac{2ab}{a} \rightarrow x = 2b.$

87. **(A)** $x = 1.$ $2x - 1 = x \rightarrow 2x - 1 + 1 = x + 1 \rightarrow 2x - x = 1 \rightarrow x = 1.$

89. **(B)** $x = \dfrac{2}{5}.$ $2x - 1 = 1 - 3x \rightarrow 2x - 1 + 1 = 1 - 3x + 1 \rightarrow 2x = 2 - 3x \rightarrow$

$$2x + 3x = 2 - 3x + 3x \rightarrow 5x = 2 \rightarrow \frac{5x}{5} = \frac{2}{5} \rightarrow x = \frac{2}{5}.$$

91. (A) $x = \dfrac{5}{11}$. $\dfrac{x-1}{3} = \dfrac{1-3x}{2} \rightarrow 3 \cdot \dfrac{x-1}{3} = 3 \cdot \dfrac{1-3x}{2} \rightarrow$

$$x - 1 = \frac{3}{2} - \frac{9}{2}x \rightarrow x = \frac{3}{2} - \frac{9}{2}x + 1 \rightarrow x + \frac{9}{2}x = \frac{5}{2} \rightarrow$$

$$\frac{11}{2}x = \frac{5}{2} \rightarrow \frac{2}{11} \cdot \frac{11}{2}x = \frac{2}{11} \cdot \frac{5}{2} \rightarrow x = \frac{5}{11}.$$

93. (D) $x = \dfrac{15}{4}$. $\dfrac{x}{2} - \dfrac{3}{4} = \dfrac{9}{8} \rightarrow \dfrac{x}{2} - \dfrac{3}{4} + \dfrac{3}{4} = \dfrac{9}{8} + \dfrac{3}{4} \rightarrow \dfrac{x}{2} = \dfrac{9}{8} + \dfrac{6}{8} \rightarrow$

$$\frac{x}{2} = \frac{15}{8} \rightarrow x = \frac{15}{4}.$$

95. (B) \$34. Let **P** represent Peter's share, **A** represent Paul's share, and **M** represent Mary's share. Then $P + A + M = 153$, $P = 2A$, $M = 3P$ Rewrite each expression in terms of Peter's share:

$P = 2A \rightarrow A = \dfrac{P}{2}$, $P + A + M = 153 \rightarrow P + \dfrac{P}{2} + 3P = 153 \rightarrow$

$$\frac{9}{2}P = 153 \rightarrow \frac{2}{9} \cdot \frac{9}{2}P = \frac{2}{9} \cdot 153 \rightarrow P = 34$$

97. (B) 7. Let **J** represent Jane's age, **B** represent Bob's age, and **I** represent

Bill's age. Then $J + B + I = 27$, $J = 2B + 3$, $I = B - 4$. $J + B + I = 27 \rightarrow (2B + 3) + B + B - 4 = 27 \rightarrow$

$$4B - 1 = 27 \rightarrow 4B = 28 \rightarrow B = 7.$$

99. (B) 36. Let **F** represent a father's current age and **S** his son's current age, where $F = 4S$. In **24** years the father's age will be $F + 24$ and his son's age will be $S + 24$, then $F + 24 = 2(S + 24) \rightarrow 4S + 24 = 2S + 48 \rightarrow 2S = 24 \rightarrow S = 12$. Therefore the son's current age is **12** years of age, and the son's age in 24 years is $12 + 24 = 36$.

101. (B) $x = -1$ or $x = -2$. $x^2 + 3x + 2 = 0 \rightarrow (x+1)(x+2) = 0 \rightarrow x = -1, x = -2$.

103. (A) $x = -1$ or $x = 2$. $x^2 - x - 2 = 0 \rightarrow (x+1)(x-2) = 0 \rightarrow x = -1, x = 2$.

105. (C) $x < -1$. $-2x > x + 3 \rightarrow \dfrac{-2}{-2}x < \dfrac{x+3}{-2} \rightarrow x < -\dfrac{x}{2} - \dfrac{3}{2} \rightarrow$

$$x + \frac{x}{2} < -\frac{x}{2} - \frac{3}{2} + \frac{x}{2} \rightarrow \frac{3x}{2} < -\frac{3}{2} \rightarrow$$

$$\frac{2}{3} \cdot \frac{3x}{2} < \frac{2}{3} \cdot \left(-\frac{3}{2}\right) \rightarrow x < -1.$$

107. (A) $x \leq -1$. $2 + 5x \leq -3 \rightarrow 2 + 5x - 2 \leq -3 - 2 \rightarrow 5x \leq -5 \rightarrow$

$$\frac{5}{5}x \leq \frac{-5}{5} \rightarrow x \leq -1.$$

109. True.

111. **False.** $x(2x+3) = 2x^2 + 3x$.

113. **(A)** $a(-1)$, **(B)** $-1 \cdot a$, **(C)** $\frac{a}{-1}$, and **(D)** $\frac{-a}{1}$.

115. **(B)** $(a-b) \div 2$, **(C)** $\frac{a}{2} - \frac{b}{2}$, **(D)** $\frac{-b+a}{2}$, and **(E)** $\frac{1}{2}(a-b)$.

117. **(A)** $(a-2b)(a-2b)$, **(D)** $a^2 - 4ab + 4b^2$, and **(E)** $-2(2b-a)(a-2b)$.

119. **(A)** $6x^3 y^6$, **(B)** $6(xy^4)^2$, and **(E)** $\frac{12x^6 y^7}{2x^3 y}$.

121. **(A)** $x \le 2$.

123. **(B)** $x \ge 65$.

125. **(A)** $-8 < x < -2$.

127. **(B)** $n \le \frac{9}{7}$. This problem can be solved using the following steps:

$$8(n-1) + 5 \ge 3(5n-4)$$
$$8n - 8 + 5 \ge 15n - 12$$
$$8n - 3 \ge 15n - 12$$
$$9 \ge 7n$$
$$\frac{9}{7} \ge n$$
$$n \le \frac{9}{7}.$$

APPENDIX D: THEOREMS, COROLLARIES, AND PROOFS

Theorem 2.1 (Cancellation Law) Let m, n, k be natural numbers with $n \neq 0$ and $k \neq 0$. Then

$$\frac{mk}{nk} = \frac{m}{n}.$$

Proof: *Prove this for $m \leq n$. For $m > n$ the argument is similar. Divide the interval $[0; 1]$ into $n\,k$ pieces. Consider the k-th, $2k$-th, $3k$-th, ..., nk-th markers. Since $\frac{nk}{nk} = 1$, the nk-th marker has to be 1. Thus the n markers k-th, $2k$-th, $3k$-th,..., nk-th, form a division of $[0; 1]$ into n equal spaces. It follows that the k-th marker is $\frac{1}{n}$, that is, $\frac{k}{nk} = \frac{1}{n}$, the $2k$-th marker is $\frac{2}{n}$, that is, $\frac{2k}{nk} = \frac{2}{n}$, etc., and so the mk-th marker is $\frac{m}{n}$, that is, $\frac{mk}{nk} = \frac{m}{n}$, as we wanted to prove.*

Thus given a fraction, if the numerator and the denominator have any common factors greater than 1, that is, any nontrivial factors, we may reduce the fraction and get an equal fraction.

Theorem 2.2 (Sum of Fractions) Let a, b, c, d be natural numbers with $b \neq 0$ and $d \neq 0$. Then

$$\frac{a}{b} + \frac{c}{d} = \frac{ad + bc}{bd}.$$

Proof: *From the Cancellation Law (Theorem 42),*

$$\frac{a}{b} + \frac{c}{d} = \frac{ad}{bd} + \frac{bc}{bd} = \frac{ad + bc}{bd},$$

proving the theorem.

The formula obtained in the preceding theorem agrees with that of (2.2) when the denominators are equal. For, using the theorem,

$$\frac{x}{b} + \frac{y}{b} = \frac{xb}{b \cdot b} + \frac{yb}{b \cdot b} = \frac{xb + by}{b \cdot b} = \frac{b(x + y)}{b \cdot b} = \frac{x + y}{b},$$

where we have used the distributive law.

Observe that the trick for adding the fractions in the preceding theorem was to convert them to fractions of the same denominator.

Theorem 2.3 (Multiplication of Fractions) Let *a*, *b*, *c*, **d** be natural numbers with $b \neq 0$ and $d \neq 0$. Then

$$\frac{a}{b} \cdot \frac{c}{d} = \frac{ac}{bd}.$$

Proof: *First consider the case when* $a = c = 1$. *Start with a unit square and cut it horizontally into* **b** *equal segments. Then cut it vertically into* **d** *equal segments. We have now* **bd** *equal pieces, each one having an area of* $\frac{1}{bd}$. *Since each piece is in dimension* $\frac{1}{b}$ *by* $\frac{1}{d}$, *we have shown that*

$$\frac{1}{b} \cdot \frac{1}{d} = \frac{1}{bd}.$$

Now construct a rectangle of length $\frac{a}{b}$ *and width* $\frac{c}{d}$. *Such a rectangle is obtained by concatenating along its length* **a** *segments of length* $\frac{1}{b}$ *and along its width* **c** *segments of length* $\frac{1}{d}$. *This partitions the large rectangle into* **ac** *subrectangles, each of area* $\frac{1}{bd}$. *Hence the area of the* $\frac{a}{b}$ *by* $\frac{c}{d}$ *rectangle is* $ab\left(\frac{1}{bd}\right)$, *from where*

$$\frac{a}{b} \cdot \frac{c}{d} = ac\left(\frac{1}{bd}\right) = \frac{ac}{bd},$$

proving the theorem.

Theorem 2.4 (Division of Fractions) Let *a*, *b*, *c*, *d* be natural numbers with $b \neq 0, c \neq 0, d \neq 0$. Then

$$\frac{a}{b} \div \frac{c}{d} = \frac{a}{b} \cdot \frac{d}{c} = \frac{ad}{bc},$$

that is, $\frac{x}{y}$ in definition 53 is $\frac{x}{y} = \frac{ad}{bc}$.

Proof: *Let us prove that* $\frac{x}{y} = \frac{ad}{bc}$ *satisfies the definition of division of fractions. Observe that*

$$\frac{x}{y} \cdot \frac{c}{d} = \frac{ad}{bc} \cdot \frac{c}{d} = \frac{adc}{bcd} = \frac{a}{b},$$

and hence $\frac{x}{y}$ *is the right result for the division of fractions. Could there be another fraction, say* $\frac{x'}{y'}$ *that satisfies definition 53? Suppose*

$$\frac{a}{b} = \frac{x'}{y'} \cdot \frac{c}{d}.$$

Then

$$\frac{x}{y} = \frac{a}{b} \cdot \frac{d}{c} = \frac{x'}{y'} \cdot \frac{c}{d} \cdot \frac{d}{c} = \frac{x'}{y'}.$$

Hence $\frac{x}{y} = \frac{a}{b} \cdot \frac{d}{c}$ *is the only fraction that satisfies the definition of fraction division.*

Theorem 4.1 (First Law of Exponents) Let a be a real number and m, n natural numbers. Then

$$a^m a^n = a^{m+n}$$

Proof: *We have*

$$a^m a^n = \underbrace{a \cdot a \cdots a}_{m \ a's} \cdot \underbrace{a \cdot a \cdots a}_{n \ a's}$$
$$= \underbrace{a \cdot a \cdots a}_{m+n \ a's}$$
$$= a^{m+n}.$$

Theorem 4.2 (Second Law of Exponents) Let $a \neq 0$ be a real number and m, n natural numbers, such that $m \geq n$. Then

$$\frac{a^m}{a^n} = a^{m-n}.$$

Proof: *We have*

$$\frac{a^m}{a^n} = \frac{\overbrace{a \cdot a \cdots a}^{m \ a's}}{\underbrace{a \cdot a \cdots a}_{n \ a's}}$$
$$= \underbrace{a \cdot a \cdots a}_{m-n \ a's}$$
$$= a^{m-n}.$$

Theorem 4.3 (Third Law of Exponents) Let a be a real number and m, n natural numbers. Then

$$(a^m)^n = a^{mn}$$

Proof: *We have*

$$(a^m)^n = \underbrace{a^m \cdot a^m \cdots a^m}_{n \ a^m's}$$
$$= a^{\overbrace{m+m+\cdots+m}^{n \ a's}}$$
$$= a^{mn}.$$

Theorem 4.4 (Fourth Law of Exponents) Let a and b be real numbers and let m be a natural number. Then

$$(ab)^m = a^m b^m$$

Proof: *We have*

$$(ab)^m = \underbrace{ab \cdot ab \cdots ab}_{m \ ab's}$$
$$= \underbrace{a \cdot a \cdots a}_{m \ a's} \underbrace{b \cdot b \cdots b}_{m \ b's}$$
$$= a^m b^m.$$

Theorem 4.5 (Square of a Sum) For all real numbers a, b the following identity holds:

$$(a \pm b)^2 = a^2 \pm 2ab + b^2$$

Proof: *Using the distributive law,*

$$
\begin{aligned}
(a + b)^2 &= (a + b)(a + b) \\
&= a(a + b) + b(a + b) \\
&= a^2 + ab + ba + b^2 \\
&= a^2 + 2ab + b^2
\end{aligned}
$$

and so $(a + b)^2 = a^2 + 2ab + b^2$. Putting $-b$ instead of b in this last identity,

$$(a - b)^2 = (a + (-b))^2 = a^2 + 2a(-b) + (-b)^2 = a^2 - 2ab + b^2$$

proving both identities.

Theorem 4.6 (Difference of Squares) For all real numbers a, b the following identity holds:

$$(a + b)(a - b) = a^2 - b^2$$

Proof: *Using the distributive law,*

$$
\begin{aligned}
(a + b)(a - b) &= a(a - b) + b(a - b) \\
&= a^2 - ab + ba - b^2 \\
&= a^2 - b^2
\end{aligned}
$$

proving the identity.

Theorem 4.7 Let a and b be real numbers. Then

$$(a + b)^3 = a^3 + 3a^2b + 3ab^2 + b^3$$

Proof: *Using the distributive law and the square of a sum identity,*

$$
\begin{aligned}
(a + b)^3 &= (a + b)^2(a + b) \\
&= (a^2 + 2ab + b^2)(a + b) \\
&= a^2(a + b) + 2ab(a + b) + b^2(a + b) \\
&= a^3 + a^2b + 2a^2b + 2ab^2 + b^2a + b^3 \\
&= a^3 + 3a^2b + 3ab^2 + b^3
\end{aligned}
$$

as claimed.

Corollary 4.1 Let a and b be real numbers. Then

$$(a - b)^3 = a^3 - 3a^2b + 3ab^2 - b^3$$

Proof: *Replace **b** by **–b** in Theorem 163, obtaining*

$$(a - b)^3 = (a + (-b))^3$$
$$= a^3 + 3a^2(-b) + 3a(-b)^2 + (-b)^3$$
$$= a^3 - 3a^2b + 3ab^2 - b^3$$

as claimed.

Theorem 4.8 (Sum and Difference of Cubes) For all real numbers **a, b** the following identity holds:

$$(a \pm b)(a^2 \mp ab + b^2) = a^3 \pm b^3$$

Proof: *Using the distributive law,*

$$(a + b)(a^2 - ab + b^2) = a^3 - a^2b + ab^2 + ba^2 - ab^2 + b^3$$
$$= a^3 + b^3$$

from where the sum of cubes identity is deduced. We obtain

$$(a - b)(a^2 + ab + b^2) = a^3 - b^3$$

*upon replacing **b** by **–b** in the above sum of cubes identity.*

$$2x = -88 \Rightarrow x = \frac{-88}{2}$$
$$\Rightarrow x = -44$$

Verification: $2x = 2(-44) \overset{\checkmark}{=} -88$

◀

◀

APPENDIX E: KEYWORDS

Addition

Additive identity

Algebra

Algebraic equation

Algebraic formula

Algebraic fraction

Algebraic generalization

Arithmetic operation

Arithmetic problem

Arithmetic progression

Axiom

Coefficient

Cumulative

Commutativity

Complete the square

Compound linear equality

Concatenation

Cube

Cubic root

Difference of cube

Difference of square

Distributive Law

Dividend

Division

Divisor

Elementary algebra

Equality

Equation

Equation of Condition

Exponential

Exponent

Exponentiation

Factor

Factoring

Factorization

Formula

Geometric progression

Greatest common divisor

Index of root

Interval

Irrational expression

Irrational number

Law of formation and conjecture

Length of interval

Linear Equation One-Variable

Magic square

Simple Equation

Multiplication

Multiplicative identity

Natural number

*n*th power

*n*th term

One–variable linear equality

One–variable linear inequality

Place-value notation system

Polynomial

Positional notation

Positive integer

Power

Product

Quadratic equation

Quadratic trinomials

Quadratic formula

Quotient

Rational expression

Rational number

Real number

Reciprocal

Recursion method

Remainder

Root

Root extraction

Special factorization

Square

Square of difference

Square of sum

Square root

Subtraction

Sum of cube

Symbol

Symbolic expression

Term-by-term division

Trice

Unknown variable

Variable

Vertex

Word Problem

INDEX